SpringerBriefs in Molecular Science

Electrical and Magnetic Properties of Atoms, Molecules, and Clusters

Series editor

George Maroulis, Patras, Greece

More information about this series at http://www.springer.com/series/11647

Akbar Salam

Non-Relativistic QED Theory of the van der Waals Dispersion Interaction

 Springer

Akbar Salam
Department of Chemistry
Wake Forest University
Winston-Salem, NC
USA

ISSN 2191-5407   ISSN 2191-5415   (electronic)
SpringerBriefs in Molecular Science
ISSN 2191-5407   ISSN 2191-5415   (electronic)
SpringerBriefs in Electrical and Magnetic Properties of Atoms, Molecules, and Clusters
ISBN 978-3-319-45604-1   ISBN 978-3-319-45606-5   (eBook)
DOI 10.1007/978-3-319-45606-5

Library of Congress Control Number: 2016949594

Printed on acid-free paper

This Springer imprint is published by Springer Nature
The registered company is Springer International Publishing AG
The registered company address is: Gewerbestrasse 11, 6330 Cham, Switzerland

To
S, R and A

And
In Memoriam: T

Preface

Why, it may reasonably be asked, write on the subject of dispersion forces and add to the already existing high-quality literature dealing with molecular QED theory? A complete answer to the question emerges after consideration of several diverse aspects. The dispersion interaction occurs between all material particles from the atomic scale upward and is a purely quantum mechanical phenomenon. Employing a quantised electromagnetic field–matter coupling approach allows for its rigorous calculation, along with an elementary understanding of the origin and manifestation of this fundamental interaction. Fluctuations of the ground state charge and current densities of the source and the vacuum field interact via the propagation of virtual photons—by definition unobservable quanta of light, resulting in an attractive force between atoms and molecules. The ubiquitous nature of the dispersion interaction means that it impacts a wide range of scientific disciplines and subareas. An opportunity therefore presents itself to bring the pioneering work of Casimir and Polder to an even broader audience, one who might ordinarily only be well versed with the London dispersion formula, by exposing them to the eponymous potential associated with the two aforementioned Dutch physicists, and the extension of their result to related applications involving contributions from higher multipole moment terms and/or coupling between three particles. This topic is also timely from the point of view that lately there has been renewed interest in a variety of van der Waals dominated processes, ranging from the physisorption of atoms and small molecules on semiconductor surfaces, to the hanging and climbing ability of geckos. These and many other problems continue to be studied experimentally and theoretically. In this second category, advances have occurred at both the microscopic and the macroscopic levels of description, frequently within the framework of QED.

Inspired by Casimir's original calculation of the force of attraction between two perfectly conducting parallel plates, much research has ensued in which the dispersion interaction has been evaluated, often within the confines of Lifshitz theory, for a plethora of different objects including plate, surface, slab, sphere, cylinder, and wedge, possessing a variety of magnetodielectric characteristics while adopting numerous geometrical configurations. In this respect, the recently published readable and comprehensive two-volume set by Stefan Buhmann titled *Dispersion Forces I* and *II* (Springer 2012) on the application of macroscopic QED to Casimir, Casimir–Polder, and van der Waals forces is recommended for its scope and detail. Similarly, with density functional theory now such a routine method that is being employed for the computation of electronic and structural properties of atomic, molecular and extended systems, prompting re-examination of elementary particle level treatments, the availability of van der Waals corrected functionals has allowed a wider class of problem to be tackled both accurately and efficiently.

This book therefore concentrates on the van der Waals dispersion interaction between atoms and molecules calculated using the techniques of molecular QED theory. Detailed presentations of this formalism may be found in the monographs published by Craig and Thirunamachandran in 1984 and by the present author in 2010. Consequently, only a brief outline of QED in the Coulomb gauge is given in Chap. 2, sufficient to understand the computations of the dispersion energy shift that follow. Evaluation of interaction energies among two and three particles, in the electric dipole approximation or beyond, is restricted to diagrammatic perturbation theory methods. This starts with a presentation of the calculation of the Casimir–Polder potential in Chap. 3. A summary is first given of its evaluation via the minimal coupling scheme, followed by its computation using the multipolar Hamiltonian. Short- and long-range forms of the interaction energy are obtained, corresponding to London and Casimir shifts, respectively. In Chap. 4, the electric dipole approximation is relaxed, and contributions to the pair potential from electric quadrupole, electric octupole, magnetic dipole, and diamagnetic coupling terms are computed. Extension to three atoms or molecules is dealt with in Chap. 5 by considering the leading non-pairwise additive contribution to the dispersion interaction, namely the triple-dipole energy shift. A retardation-corrected expression is derived first, which is shown to reduce to the Axilrod–Teller–Muto potential in the near zone. A general formula is obtained in Chap. 6 for the three-body dispersion energy shift between species possessing pure electric multipole polarisability characteristics of arbitrary order, from which is extracted specific contributions which are dependent upon combinations of dipole, quadrupole, and octupole moments valid for scalene and equilateral triangle geometries and for three particles lying in a straight line.

For those readers interested in greater detail, or alternative computational schemes, references cited at the end of each chapter may be consulted, with the caveat that the bibliography listed is far from exhaustive, with many landmark publications knowingly left out. For this choice, responsibility rests solely with the author, as with any errors that are discovered in the text.

Winston-Salem, NC, USA   Akbar Salam
June 2016

Acknowledgments

Professor George Maroulis, editor for the Springer Briefs Series in Electrical and Magnetic Properties of Atoms, Molecules and Clusters, is acknowledged for the kind invitation to contribute a volume to this series. Sonia Ojo, Esther Rentmeester, Stefan van Dijl, and Ravi Vengadachalam, production editors at Springer International Publishing AG at various stages of writing, are thanked for offering their assistance and for periodically checking in with me to ensure the manuscript remained on schedule.

Gratitude is expressed to Wake Forest University for the award of a Reynolds Research leave for Spring 2016 semester, which enabled the timely completion of this project. The contributions of Drs. Jesus Jose Aldegunde, Susana Gomez Carrasco, and Lola Gonzalez Sanchez to this endeavour cannot be overstated. Their generosity of spirit, warmth and depth of friendship, and limitless tolerance over the past four years, during my annual visits to the Departamento de Quimica Fisica, Universidad de Salamanca, especially throughout my two month stay during winter 2016, where the final three chapters were word processed, is hereby recognised and expressed. Muchas gracias amigos! These last three individuals are also thanked for many useful discussions on the topic of molecular QED. Jesus read the manuscript in its entirety and provided valuable feedback. In a similar vein, Dr. Stefan Buhmann of the University of Freiburg, Germany, is thanked for a careful and critical reading of Chap. 1. His comments and suggestions have helped to improve presentation of the background material covered in this introductory portion.

Finally, T.R. Salam and S. French are thanked for their unconditional support and continued encouragement and for their prescience in anticipating this book project well before it was even conceived.

Contents

Chapter 1
Introduction

Abstract The concept of an inter-particle potential energy is introduced, and long- and short-range interaction regions are identified. Following the multipole series expansion of the charge density, the static coupling potential between two electronic distributions is obtained. Quantum mechanical perturbation theory is then employed to extract the electrostatic, induction and dispersion energy contributions to the total interaction energy at long-range. To account for the electromagnetic nature of forces between particles of matter, the photon is introduced within the framework of quantum electrodynamics theory. Manifestations of dispersion forces between microscopic entities and macroscopic bodies are briefly reviewed.

Keywords Inter-particle potential · Quantum electrodynamics · Real and virtual photons · Van der Waals dispersion force · Casimir-Polder energy · Casimir shift

1.1 The Inter-Particle Potential

Evidence that forces operate amongst the constituent particles of a material sample is provided by the observation that matter exists in distinct physical phases [1]. With objects that are infinitely far apart taken to have zero energy of interaction, this quantity becomes increasingly negative as the entities are brought closer together, changing in curvature before reaching a minimum. As the inter-particle separation distance is further reduced, to displacements corresponding to the overlap of charge clouds, repulsive forces eventually prevail. This is confirmed by the incompressibility of condensed forms of matter, and ultimately of species in the gaseous state too, the latter situation supporting the hypothesis of finite volume associated with microscopic particles of matter. Between these two extremes of large and small separation distance, the stationary point on the energy versus coordinate profile already mentioned, corresponds to the configuration in which the two competing forces of attraction and repulsion exactly balance one another. What have been described are the characteristic features of a potential energy curve or surface, which were first recognised by Clausius [2], who heralded the concept of

© The Author(s) 2016
A. Salam, *Non-Relativistic QED Theory of the van der Waals Dispersion Interaction*, SpringerBriefs in Electrical and Magnetic Properties of Atoms, Molecules, and Clusters, DOI 10.1007/978-3-319-45606-5_1

an inter-particle force. The formation of a particular state of matter attests to the fact that attractive forces dominate overall. Commonly, the stationary point provides a convenient dividing line between short- and long-range regions, or between the repulsive and attractive components to the potential energy, $U(R)$

$$U(R) = U_{repulsive}(R) + U_{attractive}(R), \qquad (1.1)$$

where R is the inter-particle separation distance. $U(R)$ may be interpreted as an energy shift between the total energy and that of each individual species, or as the difference in energy evaluated at R and at infinity. It is often given the symbol ΔE. Clearly, the total energy is a sum of one-body, two-body, three-body, ..., many-body contributions.

Significant progress in understanding the manifestation of forces among microscopic entities, at least of particles in the gaseous state, occurred with the statistical mechanical theories of Maxwell [3] and Boltzmann [4], culminating in the kinetic molecular theory of gases. This work in turn inspired van der Waals to formulate his equation of state [5], enabling real as opposed to ideal gases to now be treated, and finally allowing phase transformations between matter to be explained. He introduced two parameters in his equation. One of these took account of the fact that the reduction in the pressure of the gas was proportional to the density due to the presence of attractive forces, while the second variable made proper allowance for the definite volume of each individual particle. The association of van der Waals' name with inter-particle forces occurring between neutral species continues to this day.

1.2 The Born-Oppenheimer Approximation

Underlying the notion of the molecular potential energy curve, and the partitioning in Eq. (1.1), is the Born-Oppenheimer approximation [6], in which the electrons adjust their positions instantaneously to any change in the electromagnetic field originating from the nuclei. The approximation is especially useful in the elucidation of molecular electronic structure [7]. It rests on the large difference in mass between protons and neutrons on the one hand, and electrons on the other. At this stage in the theoretical development, where the Coulomb interaction terms between charged particles are expressed explicitly, the nuclear kinetic energy terms are dropped from the expression for the total internal energy of the collection of charged particles that have been grouped into atoms and/or molecules. This leaves the electronic Schrödinger equation, which is then solved for a specific nuclear configuration in the clamped nuclei approximation. The electronic wave function, $\psi_{elec}[(\vec{r}_\alpha), \{\vec{R}_\xi\}]$, depends explicitly on the position of electron α, $\vec{r}_\alpha$, and similarly to the electronic energy, E_{el}, it depends parametrically on the nuclear coordinates of centre ξ, $\vec{R}_\xi$.

Solving the electronic Schrödinger equation at different nuclear positions yields E_{el}, to which is added the repulsion energy between nuclei, thereby generating the potential energy surface, $E[\{\vec{R}_\xi\}]$. Adding the nuclear kinetic energy terms to E gives rise to the Hamiltonian function for the nuclei, whose solution yields the total or molecular energy, E_{mol}, which includes electronic, vibrational, rotational and translational degrees of freedom, as well as knowledge of the nuclear wave function, $\psi_{nuc}[\{\vec{R}_\xi\}]$, which correctly accounts for the last three named possible dynamical modes. The total wave function, dependent upon electronic and nuclear coordinates, has therefore been factored into a product of nuclear and electronic wave functions, $\Psi_{tot}[(\vec{r}_\alpha), \{\vec{R}_\xi\}] = \psi_{nuc}[\{\vec{R}_\xi\}]\psi_{elec}[(\vec{r}_\alpha), \{\vec{R}_\xi\}]$ in the Born-Oppenheimer approximation.

The effects of spin may be described automatically within relativistic quantum mechanics [8], or included as an additional postulate in any non-relativistic theory, as is often the case in chemistry.

1.3 The Interaction Energy at Long-Range

Interactions between microscopic forms of matter, whether they be atoms, or electrically neutral molecules, or charged entities such as ions, are ultimately a manifestation of electromagnetic effects since each composite species is an aggregate of elementary charged particles. For separation distances between objects comprising a physical phase that are considerably larger than the dimensions of the constituent particles, it is commonplace to expand the electronic charge distribution associated with each specific centre in a multipolar series about a particular reference point. This may be the origin of charge, or the centre of mass, or an inversion centre, or the site of an individual chromophore or functional group. Often the expansion is limited to consideration of the electric polarisation field only, thereby giving rise to electric charge, dipole, quadrupole, ..., terms. Outside the region of overlap of the atomic or molecular charge distribution, the long-range interaction energy arising from the Coulomb potential between all of the electrons within each object, and between individual centres, is evaluated via Rayleigh-Schrödinger perturbation theory. Taking the expectation value of the coupling operator over the ground electronic state of each interacting particle does this. Hence the shift in energy of the interacting system relative to the energy of each isolated species when the multipolar expansion is taken to various orders in perturbation theory allows for the following decomposition to be made, namely

$$\Delta E_{long-range} = \Delta E_{elec} + \Delta E_{ind} + \Delta E_{disp}, \tag{1.2}$$

in which the interaction energy at long-range is separated into a sum of electrostatic, induction and dispersion energy contributions [9].

Explicit expressions for each of the three contributing terms are easily obtained on taking for the perturbation operator, the multipolar expansion of the potential coupling two of the species within the sample, A and B, which are separated by a distance R,

$$V_{AB} = (4\pi\varepsilon_0)^{-1} \left\{ \frac{e^A e^B}{R} + \left(e^A \mu_i(B) \nabla_i \frac{1}{R} - e^B \mu_i(A) \nabla_i \frac{1}{R} \right) - \mu_i(A) \mu_j(B) \nabla_i \nabla_j \frac{1}{R} + \cdots \right\},$$

(1.3)

where e^ξ is the net charge, and $\mu_i(\xi)$ is the electric dipole moment, of species ξ. This is the frequently encountered pairwise approximation to the total interaction energy. Equation (1.3) represents an infinite sum of familiar charge-charge, charge-electric dipole, electric dipole-electric dipole, ..., interaction terms. The Roman subscripts denote Cartesian tensor components with an implied Einstein summation convention over repeated indices. The unperturbed Hamiltonian operator is given by the sum of the Hamiltonians for the two particles,

$$H_0 = H_{part}(A) + H_{part}(B),$$

(1.4)

assuming that there is negligible exchange of electrons at long-range, so that the charged particles associated with each species remain with that particular centre. Each individual atomic or molecular Hamiltonian, a sum of kinetic and intra-molecular potential energy terms, satisfies the eigenvalue equation

$$H_{part}(\xi)|m^\xi\rangle = E_m^\xi |m^\xi\rangle, \quad \xi = A, B,$$

(1.5)

where E_m^ξ is the energy of species ξ when it is in the quantum mechanical state $|m^\xi\rangle$ corresponding to the wave function $\psi_m(\xi)$, m being the pertinent quantum number. The eigenstates of $H_{part}(\xi)$ may therefore be used as base states in the perturbation theory treatment.

1.4   Electrostatic Energy

Taking the expectation value of Eq. (1.3) over the non-degenerate ground state of each particle, $|0^A\rangle|0^B\rangle = |0^A, 0^B\rangle$, first-order perturbation theory yields the electrostatic energy,

$$\Delta E_{elec} = \langle 0^B, 0^A | V_{AB} | 0^A, 0^B \rangle$$
$$= (4\pi\varepsilon_0)^{-1} \left\{ \frac{e^A e^B}{R} + \left(e^A \mu_i^{00}(B) - e^B \mu_i^{00}(A) \right) \nabla_i \frac{1}{R} - \mu_i^{00}(A) \mu_j^{00}(B) \nabla_i \nabla_j \frac{1}{R} + \cdots \right\},$$

(1.6)

where the ground state permanent electric dipole moment of species ξ is $\mu_i^{00}(\xi) = \langle 0^\xi | \mu_i(\xi) | 0^\xi \rangle$, with similar definitions occurring for permanent moments involving higher electric multipoles. From the form of the coupling potential V_{AB}, Eq. (1.3), it is clear that ΔE_{elec} is strictly pairwise additive. The total electrostatic energy is given by summing the contributions from all pairs, with care being taken to avoid double counting of interactions. Furthermore, the energy shift (1.6) may be attractive or repulsive. When the coupling is between neutral particles, Eq. (1.6) is also known as the Keesom interaction [10], whose leading term is the static coupling between ground state permanent electric dipole moments of each species, displaying the characteristic inverse cubic dependence on separation distance,

$$\Delta E_{elec}^{\mu\mu} = \frac{\mu_i^{00}(A)\mu_j^{00}(B)}{4\pi\varepsilon_0 R^3}(\delta_{ij} - 3\hat{R}_i\hat{R}_j). \tag{1.7}$$

Higher-order contributions follow straightforwardly.

1.5 Induction Energy

Substituting the coupling operator V_{AB} into the expression for the second-order perturbative correction to the energy shift,

$$\Delta E^{(2)} = -\sum_{\substack{r,s \\ r\neq 0 \\ s\neq 0}} \frac{|\langle 0^B, 0^A | V_{AB} | 0^A, 0^B \rangle|^2}{(E_r^A - E_0^A) + (E_s^B - E_0^B)}, \tag{1.8}$$

gives rise to the second and third terms contributing to the interaction energy at long-range, namely the induction and dispersion energy shifts. In formula (1.8), the sum is taken over a complete set of excited electronic states of A and B, $|r^A\rangle$ and $|s^B\rangle$, respectively, with energies E_r^A and E_s^B. The denominators in Eq. (1.8) correspond to differences in energy between excited and ground levels in each species. It is seen that the term in which $|r^A\rangle$ and $|s^B\rangle$ are simultaneously identical to the ground electronic state is excluded from the sum over intermediate electronic states.

Considering the case in which A is electronically excited, but B remains in the lowest level, produces

$$\Delta E_{ind}(A) = -\sum_{r\neq 0} \frac{\langle 0^B, 0^A | V_{AB} | r^A, 0^B \rangle \langle 0^B, r^A | V_{AB} | 0^A, 0^B \rangle}{E_r^A - E_0^A}$$

$$= -\frac{1}{2}\left(\frac{1}{4\pi\varepsilon_0}\right)^2 \left[e^B \nabla_i \frac{1}{R} - \mu_j^{00}(B)\nabla_i\nabla_j\frac{1}{R} + \cdots\right]\alpha_{ii'}(A;0)\left[e^B\nabla_{i'}\frac{1}{R} - \mu_{j'}^{00}(B)\nabla_{i'}\nabla_{j'}\frac{1}{R} + \cdots\right], \tag{1.9}$$

which is identified as the induction energy of species A. It has a simple physical origin, arising from the static electric field due to the charge and ground state permanent moments associated with the charge distribution of B, which polarises the charge cloud around A, inducing electronic moments in the latter particle. The inducing and induced moments interact, producing the induction energy, which is always attractive. Noticing that the second equality in Eq. (1.9) may be rewritten to leading order as

$$\Delta E_{ind}(A) = -\frac{1}{2}\alpha_{ii'}(A;0)E_i(B;0)E_{i'}(B;0) + \cdots, \qquad (1.10)$$

readily makes apparent the interpretation of the induction energy as arising from the response of A, through its static electric dipole polarisability tensor,

$$\alpha_{ij}(A;0) = \sum_{r\neq 0} \frac{\mu_i^{0r}(A)\mu_j^{r0}(A) + \mu_j^{0r}(A)\mu_i^{r0}(A)}{E_r^A - E_0^A}, \qquad (1.11)$$

to the static electric field of species B,

$$E_i(B;0) = \frac{1}{4\pi\varepsilon_0}\left[e^B\nabla_i\frac{1}{R} - \mu_j^{00}(B)\nabla_i\nabla_j\frac{1}{R} + Q_{jk}^{00}(B)\nabla_i\nabla_j\nabla_k\frac{1}{R} - \cdots\right]. \qquad (1.12)$$

Appearing in the polarisability of A is the transition electric dipole moment matrix element between states $|0^A\rangle$ and $|r^A\rangle$, $\vec{\mu}^{0r}(A) = \langle 0^A|\vec{\mu}(A)|r^A\rangle$. Higher-order multipole contributions, such as the electric quadrupole dependent term, $Q_{jk}^{00}(B)$, and cubic, quartic, ..., dependencies on the static electric field ensue from Eq. (1.10) on further expansion of the charge density. Taking species A to only be in its ground state, but permitting B to be electronically excited, so that $s \neq 0$ in Eq. (1.8), gives rise to the induction energy of B. Expressions similar to Eqs. (1.9) and (1.10) follow on taking advantage of the symmetry inherent in this phenomenon, and which may be realised on effecting appropriate change of labels identifying the respective molecular state and species involved. Unlike ΔE_{elec}, the induction energy is non-additive because the electric field originating from particle A say, is not due solely to this entity, but also contains contributions from the presence of species B, as well as from other bodies C, D, ..., that comprise the sample. Often, the force between neutral species resulting from the induction energy is called the Debye force [11].

1.6 Dispersion Energy

Returning to expression (1.8), we now consider the case in which both A and B may be simultaneously excited as a result of transitions to virtual electronic levels. Thus

$$\Delta E_{disp} = -\sum_{\substack{r\neq 0 \\ s\neq 0}} \frac{\langle 0^A, 0^B \mid V_{AB} \mid s^B, r^A \rangle \langle r^A, s^B \mid V_{AB} \mid 0^B, 0^A \rangle}{(E_r^A - E_0^A) + (E_s^B - E_0^B)}, \qquad (1.13)$$

and which is identified as the dispersion energy shift. For electrically neutral systems that are non-polar, the latter implying an absence of permanent moments, Eq. (1.13) is the only term contributing to the interaction energy at long-range. Inserting the perturbation operator (1.3) into (1.13), the leading contribution to the dispersion potential between species that are electrically uncharged overall, is

$$\Delta E_{disp}^{\mu\mu} = -\frac{1}{24\pi^2 \varepsilon_0^2 R^6} \sum_{\substack{r\neq 0 \\ s\neq 0}} \frac{|\vec{\mu}^{0r}(A)|^2 |\vec{\mu}^{0s}(B)|^2}{E_{r0}^A + E_{s0}^B}, \qquad (1.14)$$

for freely tumbling A and B, with energy denominator differences written as $E_t^\xi - E_0^\xi = E_{t0}^\xi$. Result (1.14) is immediately recognisable as the London dispersion energy [12], with characteristic inverse sixth power separation distance dependence, a consequence of the static dipolar coupling operator being evaluated at second-order of perturbation theory.

The dispersion force is interpreted as arising from the coupling of transition multipole moments induced at one centre by the fluctuations in electronic charge density occurring in a second atom or molecule due to the continuous motion of electrons in that species, and vice versa. Hence Eq. (1.14), and its higher-order corrections, is often called the induced dipole (multipole)—induced dipole (multipole) interaction. It is purely quantum mechanical in origin and manifestation, having no classical analogue. The dispersion force is therefore an ever-present feature between interacting particles of all types, charged or neutral, polar or non-polar. The collection of charged particles associated with a particular atom or molecule may be considered as a source of electromagnetic radiation, whose oscillating fields polarise a second body, inducing a transient moment in it. A similar moment is induced in the first particle due to the radiation field of the second. The coupling of the temporary electric dipoles or higher multipoles produces the dispersion force as before, but which may now be viewed in terms of oscillating charges giving rise to propagating electromagnetic radiation [13]. Because the coupling between particles in the calculation by London is static rather than dynamic, the force is transmitted instantaneously, clearly an unphysical phenomenon and an artefact of only describing the material entities quantum mechanically. While causal Maxwell fields may be employed in a semi-classical treatment, the effects of retardation are rigorously accounted for through photon propagation by adopting a fully quantized theory of radiation-molecule interaction, as exemplified by molecular QED theory [14–16]. Application of the latter formalism to the study of van der Waals dispersion forces forms the subject of this brief volume.

1.7 Photons: Real and Virtual Light Quanta

A condensed outline of the construction of the theory of molecular QED will be given in Chap. 2. Its characteristic feature is that both matter and electromagnetic radiation are subject to quantum mechanical principles. Photons—elementary excitations of the radiation field, emerge automatically on quantisation [17]. Coupling of atomic and molecular charge and current distributions to the Maxwell fields may be conveniently expressed in terms of electric, magnetic and diamagnetic distributions. These may be expanded to give the familiar electric and magnetic multipole moments [18]. This proves advantageous on application of the molecular QED formalism to systems of general interest in chemical physics, as well as to the specific consideration of explicit contributions to the van der Waals dispersion energy shift to be detailed in the subsequent chapters of this book. Hence molecular QED theory not only enables inter-particle interactions to be correctly understood and evaluated, but also allows linear and nonlinear spectroscopic processes involving the single- and multi-photon absorption, emission and scattering of light to be treated accurately and consistently [19, 20].

Underlying this wide-ranging applicability of the formalism, are two types of light quanta permitted by the theory: *real* photons, which are ultimately detected, and their *virtual* counterparts, which by definition are not observed [21]. Curiously, this second type of photon may be emitted and absorbed by the same centre, thereby giving rise to a virtual cloud [22], and radiative corrections to the self-energy of a particle such as the electron, for instance [23, 24]. Virtual photons may also be exchanged between multiple sites [25]. This is the origin of coupling between material particles through the electromagnetic force.

Even though the theory of molecular QED may be derived from first principles by starting from the classical Lagrangian function for the total system comprised of a collection of charged particles, the electromagnetic field, and their mutual interaction, and carrying out the canonical quantization scheme to arrive at a QED Hamiltonian operator [14–16], it also follows on approximating the speed of electrons in atoms and molecules to be significantly less than that of light within the fully covariant formulation of QED developed by Feynman, Schwinger, Tomonaga and Dyson [26]. In this sense, molecular QED is synonymous with non-relativistic QED [27], and applies to bound electrons possessing energies $E \ll m_e c^2$, where m_e is the mass of the electron, and c is the speed of light in vacuum. In both versions, the electromagnetic field is Lorentz invariant.

Over the years molecular QED theory has been developed and applied successfully to problems occurring in atomic, molecular and optical physics, and theoretical chemistry [14–16, 19, 20]. These include a vast array of spectroscopic techniques involving one- and two-photon absorption of linearly and circularly polarised light, to forward and non-forward elastic and inelastic light scattering (Rayleigh and Raman) processes, to more sophisticated laser-matter phenomena such as sum- and difference-frequency and harmonic generation, and other effects arising from multi-wave mixing. Apart from being able to treat electron-photon interactions taking place at a single site, QED theory also permits coupling between two or more particles to be mediated by the exchange of one or more virtual photons and to be calculated using the same theoretical formalism that describes the emission and absorption of real photons. It is in this aspect of the theory that the power and versatility of QED resides. Numerous examples of fundamental inter-particle processes adeptly tackled via the techniques of a fully quantised prescription exist [15, 16]. They include resonant transfer of excitation energy, and the electrostatic energy Eq. (1.6), both of which are understood in terms of single virtual photon exchange; and the retarded dispersion potential, which arises from the exchange of two virtual photons.

Calculation of the Casimir -Polder energy shift [28], and the dispersion inter-action among three-bodies [29, 30], as well as higher-order multipole moment corrections, will be carried out using diagrammatic perturbation theory [31] in the chapters to follow. This visual depiction of electron-photon coupling events in space-time was originally devised by Feynman [32, 33]. It is a powerful aid in computing matrix elements since each diagram corresponds to a unique contributory term in the perturbation sum and series, in addition to providing a pictorial representation of the specific process under investigation.

1.8 Dispersion Forces Between Macroscopic Objects

Since all quantum field theory computations of the dispersion force necessarily involve taking the expectation value of the initial and final state of the radiation field in which there are no photons present, this interaction is often interpreted as a manifestation of fluctuations of the vacuum electromagnetic field and its associated zero-point energy [34]. In fact this idea was central to Casimir's pioneering calculation in 1948 of the attraction of two parallel conducting plates separated by a distance R [35]. He found that the force of this isolated system varied as the inverse fourth power of R. The presence of the plates served to impose boundary conditions on the vacuum electromagnetic field, resulting in the number of allowed modes between the walls being restricted relative to those outside, causing net attraction of the two surfaces. This work inspired much study, being classified under the headings of "Casimir effect" or "Casimir physics" [36–40].

Fairly soon after the appearance of Casimir's, and Casimir and Polder's work, and its extension to cover the force between two dielectrics as in the general theory formulated by Lifshitz [41], attempts were made at rederiving the various force laws using differing physical viewpoints and computational schemes within the multi-polar formalism of QED. One fruitful method, in which there was no reference made to the vacuum field and zero-point energy, was the source theory approach developed by Schwinger and co-workers in 1978 [42], and which was later applied by Milonni to dispersion forces between atoms and molecules [43, 44]. In this viewpoint, the presence of a second neutral polarisable species causes a change in the modes of the radiation field. These are composed of vacuum modes plus the dipole field created by the source itself. This in turn alters the radiation reaction field—the interaction of the particle with the field it generates, of the first particle from its free-space functional form. In the case of atoms and molecules, that part of the ground state shift in energy levels, which depends on the pair separation distance, is equal to the van der Waals dispersion energy. Both perspectives, namely that the Casimir force, and its formulation in terms of Lifshitz theory, may be attributed to the effect of the vacuum only, or due to the sources only, are legitimate and equally valid viewpoints within the confines of QED theory, and are in fact necessary for its internal consistency, ultimately resting on the property that the equal time particle and field operators commute [34]. Hence, in general in a normal ordering of field operators, in which the boson annihilation operator occurs on the right and the creation operator appears on the left in any expression involving products of operators, there is no explicit contribution from the vacuum field, with only the source field contributing. Conversely, if the field operators are ordered symmetrically, only the vacuum field contributes explicitly to the energy shift or force, with the source field playing no role. Interestingly, in a non-normal ordering of field operators, explicit contributions from both vacuum and source fields appear.

It may therefore be concluded that in QED it is the fluctuation of both charges and the electromagnetic field that gives rise to Casimir, Casimir-Polder, and van der Waals dispersion forces of attraction. One consistent way in which these phenomena have been treated when they take place in a magnetodielectric medium is by the methods of macroscopic QED [45], recovering where appropriate, pertinent results usually obtained via normal mode QED. The key feature in the construction of macroscopic QED is that a linear, isotropic medium, together with the electromagnetic field, is quantized, resulting in a body-assisted radiation field. The medium is taken to be dispersing and absorbing. It is characterised by the complex, scalar, causal, response functions $\varepsilon(\vec{r}, \omega)$, and $\mu(\vec{r}, \omega)$, the electric permittivity and magnetic permeability, respectively, that each depend upon the field point position vector, $\vec{r}$, and the circular frequency, ω. Additional properties of the medium include the random noise polarisation and magnetisation fields, $\vec{P}_N(\vec{r}, t)$ and $\vec{M}_N(\vec{r}, t)$, respectively, whose ground state fluctuations are related to the imaginary parts of the permittivity and permeability via the fluctuation-dissipation theorem. Constitutive relations allow the auxiliary fields to be defined, with the electric displacement field, $\vec{D}(\vec{r}, t)$, related to $\vec{P}_N(\vec{r}, t)$ and the fundamental electric field $\vec{E}(\vec{r}, t)$, while the magnetic field, $\vec{H}(\vec{r}, t)$, is expressed in terms of the magnetic induction field, $\vec{B}(\vec{r}, t)$, and $\vec{M}_N(\vec{r}, t)$ [45]. The fundamental and auxiliary fields obey the macroscopic Maxwell equations. Solutions for $\vec{E}(\vec{r}, t)$ and $\vec{B}(\vec{r}, t)$ are written in terms of the classical Green's tensor for an absorbing magnetodielectric medium. Second-quantised body-assisted boson annihilation and creation operators are introduced to describe photon absorption and emission events. These operators are subject to equal time commutation relations, which are analogous to those occurring in the quantisation of the free electromagnetic field. Next the body-assisted field operators are used to construct the Hamiltonian operator for the radiation field and the magnetodielectric medium. Coupling to material particles and objects enables the total system Hamiltonian function to be obtained in either the minimal- or multipolar-coupling schemes.

1.9   Different Physical Ways of Understanding the Dispersion Interaction Between Atoms and Molecules

Early methods focussing on the re-calculation of the Casimir-Polder dispersion potential between two atoms involved the use of relativistic quantum field theory and the S-matrix technique, from which the non-relativistic form with respect to the particle velocities and energies was then extracted [46–50]. Closely related to these approaches were the scattering theory calculations of Feinberg and Sucher [51]. They evaluated the amplitude for scattering of two-photons from each centre, and employed real form factor structure functions that were directly related to atomic dipole polarisability tensors. Their method also allowed for higher multipole moment contributions to be calculated [52], although care is needed when expanding the photon momentum vectors featuring in the electric and magnetic form factors and their proper identification with higher multipole moments [53, 54]. Nonetheless, their efforts resulted in the magnetic dipole analogue to the Casimir-Polder potential, as well as the energy shift between an electric dipole polarisable atom and a paramagnetically susceptible one to be calculated, with the sum of retarded dipolar dispersion interaction contributions often being referred to as Feinberg-Sucher potentials [51] . Specific higher multipolar terms to the total pair dispersion energy are presented in Chap. 4, while contributions beyond the electric dipole approximation to the three-body dispersion potential may be found in Chap. 6.

The methods described in relation to the evaluation of the Casimir effect, along with Casimir and Polder's original calculation of the dispersion force between two atoms, still necessitates the use of sophisticated mathematical techniques in technically demanding computations. Alternate approaches were sought in which the retardation corrected form of the London dispersion formula in particular, could be arrived at more simply. This challenge helped spur the direct use of non-relativistic QED theory, culminating in a wide variety of calculational schemes now being available within the framework of the multipolar formalism of molecular QED theory.

One of the most versatile methods is response theory [55–58]. In this approach, the charged particles that form atoms and molecules are viewed as a source of electromagnetic radiation. The quantum electrodynamical Maxwell field operators in the vicinity of the source are evaluated, and expanded in a series of powers of the multipole moments. Calculation of the radiation field operators is most conveniently carried out in the Heisenberg picture of quantum mechanics [8], which has the additional advantage that the ensuing formula for the quantum mechanical expression for the interaction energy operator is formally analogous to its classical counterpart. A second, electrically polarisable or magnetically susceptible atom or molecule is then taken as a test species, responding to the electric and magnetic fields of the first particle. The symmetry underlying the problem enables the roles of source and test bodies to be freely interchanged. Taking the expectation value of the operator expression for the interaction energy over the ground electronic state of each species, and the electromagnetic field in the vacuum state, yields the Casimir-Polder dispersion potential when both source and test particles are restricted to the electric dipole approximation.

Reminiscent of the way in which the Casimir force between two parallel conducting plates in vacuum was computed, the Casimir-Polder dispersion potential between two atoms was rederived by calculating the difference in zero-point energy between the electromagnetic vacuum, and a vacuum containing two ground state atoms. The presence of matter changes the mode structure of the vacuum relative to its absence. While the mode sums in each of the two situations are infinite, their difference produces a finite, measureable inter-atomic energy shift [59].

Another physically insightful method relies on the well-known phenomenon of an electric dipole being induced, to leading order, in an electrically polarisable species when subject to an applied electric field. The dipoles induced at each centre are coupled through an interaction tensor, whose static form was given by Eq. (1.7). Its retarded version is well known, featuring in the matrix element for resonant transfer of excitation energy [15, 16],

$$V_{ij}(k, \vec{R}) = \frac{1}{4\pi\varepsilon_0 R^3} [(\delta_{ij} - 3\widehat{R}_i\widehat{R}_j)(1 - ikR) - (\delta_{ij} - \widehat{R}_i\widehat{R}_j)k^2 R^2]e^{ikR}, \qquad (1.15)$$

where k is the magnitude of the wave vector of the exchanged photon. As expected, the dispersion energy shift results on taking the expectation value over the ground state of the total system of the product of dipoles induced at each site with the dipole-dipole coupling tensor (1.15) [13]. This particular approach is similar to that originally adopted by London in his solution to the problem, in which the dispersion force was interpreted as arising from the coupling of fluctuating electric dipole moments at the position of each particle [12, 60].

A final, fundamental way of interpreting and calculating the dispersion energy shift that is important to mention involves the dressed atom approach. In this particular context the term "dressed" refers to the cloud of virtual photons surrounding a bare atom that is induced by the quantum mechanical fluctuations of the vacuum electromagnetic field [61]. This results in the continuous emission and re-absorption of photons by the same centre, producing the self-energy interaction. This process is ultimately responsible, through renormalisation, of the observed charge and mass of the electron [26]. The interaction of one atom, possessing its own virtual photon field, with a second dressed atom, imbued with its own virtual cloud, between which photons are exchanged, leads to coupling between particles and the Casimir-Polder dispersion potential [28]. Interestingly, this picture bears fairly close resemblance to the response method described earlier, where in the present case a test polarisable species is coupled to the electromagnetic energy density due to the virtual photon cloud of the dressed source.

Common to all of the various approaches briefly described above to arrive at the retarded dispersion force between a pair of atoms or molecules is the concept of the photon [62], the particle that results on quantisation of the electromagnetic field, and which is an inherent feature of QED theory. Mediation of the interaction takes place through the exchange of virtual photons. When electromagnetic influences are described in terms of photons propagating with finite velocity appropriate to the medium in question, the ensuing delay between the action of the Maxwell field of the first particle, polarising and inducing moments in the second species, whose electromagnetic field couples to the first object, is $2R/c$, meaning that the induced moments are no longer oscillating in phase, thereby causing a weakening of the London dispersion potential, whose functional form is replaced by the R^{-7} dependent Casimir-Polder energy shift. Computation of the latter is given in Chap. 3. It will be seen that both the London and Casimir-Polder interaction energies may be written in terms of molecular polarisability functions [25]. This last quantity is in general complex, with the real part being related to the variation of the refractive index with frequency, while the imaginary part describes light absorption, making clear the association of the adjective dispersion with the manifestation of the force between neutral polarisable bodies. When the interacting species are in their ground electronic states, the dispersion potential is always attractive. If one or both species is electronically excited, however, the energy shift may be of either sign, and therefore has the possibility of being repulsive.

While the dominant contribution to ΔE_{disp} is obtained by summing over contributions arising from interactions between pairs of particles, the dispersion force, like the induction energy, is non-additive, with many-body correction terms coming from three-, four-, ... body contributions. In the case of dispersion energy between three particles, the sign of the potential depends on the geometry adopted by the objects in question. Perturbation theory calculation of the triple dipole dispersion energy shift is presented in Chap. 5, with contributions from higher multipole moments between pairs of molecules and three-bodies extracted in Chaps. 4 and 6, respectively. In the three-particle case, the potentials are obtained for scalene and equilateral triangles, and for a collinear configuration. As for two-body potentials, near- and far-zone asymptotically limiting forms are found for three coupled species.

A few remarks regarding the naming and classification of the various types of forces and couplings contributing to the total interaction energy of a collection of particles are in order. In short, there is no general consensus on terminology. Researchers within the same field and different sub-area, as well as across diverse disciplines have adopted a variety of definitions to categorise the physical effects in play. In this work we employ the following descriptors and naming conventions. Long-range inter-particle forces between neutral species in the ground state are termed as being of the van der Waals type, and may be followed by electrostatic, induction or dispersion as appropriate since van der Waals forces include all three of these types of interaction, but rules out the resonant interaction, even though dispersion forces occur between species in electronically excited states. Because the retarded dispersion potential between two ground state atoms first appeared in the paper published by Casimir and Polder [28], these two names are also associated with, and will refer to, the dispersion energy shift between two neutral particles. Note that in this same article, these two authors also obtained the dispersion potential between an atom and a semi-infinite half-space, which may be thought of as an atom-body interaction, the body being considered as a macroscopic object. In the literature, this interaction is also referred to as a Casimir-Polder dispersion energy shift or force. We will only be concerned with the dispersion potential between two and three microscopic particles—atoms or molecules in this work. In a similar vein, the term Casimir forces commonly describes the dispersion interaction between two semi-infinite half-spaces or bodies. A variety of shapes have been studied within the context of the last two categories defined, including surface, slab, cylinder, sphere, and wedge, along with cavities of differing geometry, with the materials themselves possessing a wealth of individual linear response characteristics such as metallic, dielectric or magnetodielectric [38, 45, 47]. Frequently, especially in the older literature, the prefixes van der Waals and Casimir-Polder were used to distinguish between non-retarded and retarded dispersion forces, respectively, with the compounded form London-van der Waals dispersion force, for example, having widespread usage to describe the situation in which coupling between centres is electrostatic. Other naming conventions will be mentioned as and when they are encountered.

References

1. Margenau H, Kestner NR (1969) Theory of intermolecular forces. Pergamon, Oxford
2. Clausius R (1857) Über die art der bewegung. Ann der Phys 100:353
3. Maxwell JC (1867) On the dynamical theory of gases. Philos Trans Roy Soc London 157:49
4. Boltzmann L (1872) Weitere studien über das wärmegleichgewicht unter gasmolekülen. Sitz Akad Wiss Wien 66:275
5. van der Waals JD (1873) On the continuity of the gas and liquid state. Doctoral dissertation, Leiden University
6. Born M, Oppenheimer JR (1927) Zur quantentheorie der molekeln. Ann Phys 84:457
7. Szabo A, Ostlund NS (1996) Modern quantum chemistry. Dover, New York
8. Dirac PAM (1958) The principles of quantum mechanics. Clarendon, Oxford
9. Buckingham AD (1967) Permanent and induced molecular moments and long-range intermolecular forces. Adv Chem Phys 12:107
10. Keesom WH (1915) The second virial coefficient for rigid spherical molecules whose mutual attraction is equivalent to that of a quadruplet placed at its centre. Proc K Ned Akad Wet 18:636
11. Debye P (1920) Die van der Waalsschen Kohäsionskräfte. Phys Z 21:178
12. London F (1930) Zur Theorie und Systematik der Molekularkräfte. Z Phys 63:245

13. Power EA, Thirunamachandran T (1993) Casimir-Polder potential as an interaction between induced dipoles. Phys Rev A 48:4761
14. Power EA (1964) Introductory quantum electrodynamics. Longmans, London
15. Craig DP, Thirunamachandran T (1998) Molecular quantum electrodynamics. Dover, New York
16. Salam A (2010) Molecular quantum electrodynamics. Wiley, Hoboken
17. Andrews DL (2013) Physicality of the photon. J Phys Chem Lett 4:3878
18. Jackson JD (1975) Classical electrodynamics. Wiley, New York
19. Mukamel S (1995) Principles of nonlinear optical spectroscopy. Oxford University Press, New York
20. Andrews DL, Allcock P (2002) Optical harmonics in molecular systems. Wiley, Wienheim
21. Andrews DL, Bradshaw DS (2014) The role of virtual photons in nanoscale photonics. Ann der Physik 526:173
22. Compagno G, Passante R, Persico F (1995) Atom-field interactions and dressed atoms. Cambridge University Press, Cambridge
23. Lamb WE Jr, Retherford RC (1947) Fine structure of the hydrogen atom by a microwave method. Phys Rev 72:241
24. Bethe HA (1947) The electromagnetic shift of energy levels. Phys Rev 72:339
25. Salam A (2015) Virtual photon exchange, intermolecular interactions and optical response functions. Mol Phys 113:3645
26. Schwinger JS (ed) (1958) Selected papers on quantum electrodynamics. Dover, New York
27. Healy WP (1982) Non-relativistic quantum electrodynamics. Academic Press, London
28. Casimir HBG, Polder D (1948) The influence of retardation on the London van der Waals forces. Phys Rev 73:360
29. Axilrod BM, Teller E (1943) Interaction of the van der Waals type between three atoms. J Chem Phys 11:299
30. Aub MR, Zienau S (1960) Studies on the retarded interaction between neutral atoms. I. Three-body London-van der Waals interaction between neutral atoms. Proc Roy Soc London A257:464
31. Ward JF (1965) Calculation of nonlinear optical susceptibilities using diagrammatic perturbation theory. Rev Mod Phys 37:1
32. Feynman RP (1949) The theory of positrons. Phys Rev 76:749
33. Feynman RP (1949) Space-time approach to quantum electrodynamics. Phys Rev 76:769
34. Milonni PW (1994) The quantum vacuum. Academic Press, San Diego
35. Casimir HBG (1948) On the attraction between two perfectly conducting plates. Proc K Ned Akad Wet B51:793
36. Milton KA (2004) The Casimir effect: Recent controversies and progress. J Phys A: Math Gen 37:R209
37. Buhmann SY, Welsch D-G (2007) Dispersion forces in macroscopic quantum electrodynamics. Prog Quantum Electron 31:51
38. Bordag M, Klimchitskaya GL, Mohideen U, Mostepanenko VM (2009) Advances in the Casimir effect. Oxford University Press, Oxford
39. Babb JF (2010) Casimir effects in atomic, molecular, and optical physics. Adv At Mol Opt Phys 59:1
40. Dalvit D, Milonni PW, Roberts D, da Rosa F (eds) (2011) Casimir physics, Lecture Notes in Physics, vol 834. Springer, Berlin
41. Lifshitz EM (1956) The molecular theory of attractive forces between solids. Sov Phys JETP 2:73
42. Schwinger JS, DeRaad LL Jr, Milton KA (1978) Casimir effect in dielectrics. Ann Phys (NY) 115:1
43. Milonni PW (1982) Casimir forces without the vacuum radiation field. Phys Rev A 25:1315
44. Milonni PW, Shih M-L (1992) Source theory of Casimir force. Phys Rev A 45:4241
45. Buhmann SY (2012) Dispersion forces I and II. Springer, Berlin

46. Dzyaloshinskii IE (1957) Account of retardation in the interaction of neutral atoms. Sov Phys JETP 3:977
47. Dzyaloshinskii IE, Lifshitz EM, Pitaevskii LP (1961) General theory of van der Waals' forces. Sov Phys Usp 4:153
48. Mavroyannis C, Stephen MJ (1962) Dispersion forces. Mol Phys 5:629
49. McLachlan AD (1963) Retarded dispersion forces between molecules. Proc Roy Soc London A271:387
50. Boyer TH (1969) Recalculations of long-range van der Waals potentials. Phys Rev 180:19
51. Feinberg G, Sucher J (1970) General theory of the van der Waals interaction: A model-independent approach. Phys Rev A 2:2395
52. Au C-KE, Feinberg G (1972) Higher-multipole contributions to the retarded van der Waals potential. Phys Rev A 6:2433
53. Power EA, Thirunamachandran T (1996) Dispersion interactions between atoms involving electric quadrupole polarisabilities. Phys Rev A 53:1567
54. Salam A, Thirunamachandran T (1996) A new generalisation of the Casimir-Polder potential to higher electric multipole polarisabilities. J Chem Phys 104:5094
55. Power EA, Thirunamachandran T (1983) Quantum electrodynamics with non-relativistic sources. II. Maxwell fields in the vicinity of a molecule. Phys Rev A 28:2663
56. Power EA, Thirunamachandran T (1983) Quantum electrodynamics with non-relativistic sources. III. Intermolecular interactions. Phys Rev A 28:2671
57. Thirunamachandran T (1988) Vacuum fluctuations and intermolecular interactions. Phys Scr T21:123
58. Salam A (2008) Molecular quantum electrodynamics in the Heisenberg picture: a field theoretic viewpoint. Int Rev Phys Chem 27:405
59. Power EA, Thirunamachandran T (1994) Zero-point energy differences and many-body dispersion forces. Phys Rev A 50:3929
60. Spruch L, Kelsey EJ (1978) Vacuum fluctuation and retardation effects on long-range potentials. Phys Rev A 18:845
61. Compagno G, Passante R, Persico F (1983) The role of the cloud of virtual photons in the shift of the ground-state energy of a hydrogen atom. Phys Lett A 98:253
62. Andrews DL (ed) (2015) Fundamentals of photonics and physics. Wiley, Hoboken

Chapter 2
Non-relativistic QED

Abstract A brief presentation is given of the construction of the theory of molecular QED. This is done by first writing a classical Lagrangian function for a collection of non-relativistic charged particles coupled to an electromagnetic field. After selecting the Coulomb gauge, Hamilton's principle is invoked and the Lagrangian is substituted into the Euler-Lagrange equations of motion and shown to lead to the correct dynamical equations. These are Newton's second law of motion with added Lorentz force law electric and magnetic field dependent terms, and the wave equation for the vector potential in the presence of sources. Canonically conjugate particle and field momenta are then evaluated, from which the Hamiltonian is derived. Elevation of classical variables to quantum operators finally yields the molecular QED Hamiltonian, which is expressed in minimal-coupling and multipolar forms. In the QED formulation, the electromagnetic field is described as a set of independent simple harmonic oscillators. Elementary excitations of the field, the photons, emerge automatically on quantisation.

Keywords Lagrangian · Polarisation · Magnetisation · Minimal-coupling Hamiltonian · Canonical transformation · Multipolar Hamiltonian · Perturbation theory

2.1 Classical Mechanics and Electrodynamics

As implied by its name, quantum electrodynamics (QED) [1] concerns the quantum mechanical description of charged particles in motion. Newton's Laws of Motion adequately treat the vast majority of kinematical situations encountered by macroscopic objects [2]. If these bodies move with velocities that are appreciable relative to that of light, however, then a Lorentz transformation may be applied, resulting in a relativistic treatment of the dynamics, as formulated by Einstein in his Special and General Theories of Relativity. A particularly elegant and advantageous form of classical mechanics, from the viewpoint of development of a quantum mechanical theory [3], is through the utilisation of the Lagrangian function, L,

© The Author(s) 2016

A. Salam, *Non-Relativistic QED Theory of the van der Waals Dispersion Interaction*, SpringerBriefs in Electrical and Magnetic Properties of Atoms, Molecules, and Clusters, DOI 10.1007/978-3-319-45606-5_2

which is defined for a conservative system as the difference between kinetic and potential energies, $T - V$, along with the invocation of Hamilton's principle. This states that the path taken by an object as it moves in configuration space from space-time point (q_1, t_1) to (q_2, t_2), where q is the generalised coordinate and t is the time, is the one for which the action, S, is a variational minimum. S is defined as the time integral of L, so that the extremum condition is

$$\delta S = \delta \int_{t_1}^{t_2} L(q, \dot{q}, t) dt = 0, \qquad (2.1)$$

where δ denotes the variation, and the velocity, $\dot{q} = dq/dt$. Standard calculus of variations [4] leads to the Euler-Lagrange equations of motion

$$\frac{d}{dt}\left(\frac{\partial L}{\partial \dot{q}_\alpha}\right) - \frac{\partial L}{\partial q_\alpha} = 0, \qquad \alpha = 1, 2, \ldots, N, \qquad (2.2)$$

for a system with N degrees of freedom. Selection of a suitable coordinate system for a specific dynamical problem often prevents the wider exploitation of Eq. (2.2) in classical mechanics when compared to applications of Newton's Second Law of Motion.

Since charged particles, protons and electrons are constituents of all matter—from the atoms of elements to the chemical compounds they form, any theory of the dynamics of such sources must also include a correct description of the electromagnetic fields that necessarily ensue, or which may be applied. These are expressed beautifully by Maxwell's equations, which encapsulate the properties and behaviour of all electromagnetic phenomena [5]. Hence classical mechanics and classical electrodynamics form two key ingredients in any theory of radiation-matter interaction. Unfortunately, these classical laws do not apply to microscopic entities. Elementary particles are instead governed according to quantum mechanical principles. This is the third, and obviously the most crucial element in the construction of QED theory. Slow or fast moving sub-atomic species, with energies that are significantly less than or comparable to mc^2, where m is the mass and c is the speed of light, may then be appropriately tackled by using non-relativistic or relativistic formulations of quantum mechanics, respectively.

To facilitate the application of quantum mechanical rules to the coupled charged particle-electromagnetic field system, Maxwell's equations are written in their microscopic form:

$$\varepsilon_0 \text{div}\vec{e} = \rho \qquad (2.3)$$

$$\text{div}\vec{b} = 0 \qquad (2.4)$$

$$\mathrm{curl}\vec{e} = -\frac{\partial \vec{b}}{\partial t} \tag{2.5}$$

$$\mathrm{curl}\vec{b} = \frac{1}{c^2}\frac{\partial \vec{e}}{\partial t} + \frac{1}{\varepsilon_0 c^2}\vec{j}, \tag{2.6}$$

with ε_0 the permittivity of the vacuum, which along with the permeability of the vacuum, μ_0, are related to the speed of light via $c^2 = (\varepsilon_0\mu_0)^{-1}$.

Maxwell's equations may be converted to their more familiar macroscopic counterparts after performing a spatial average over the fundamental microscopic electric field $\vec{e}(\vec{r}, t)$, and the magnetic induction field $\vec{b}(\vec{r}, t)$, which are both functions of position $\vec{r}$, and time, t. These fields are related to the sources via the charge and current densities, $\rho(\vec{r})$ and $\vec{j}(\vec{r})$, respectively. In a microscopic description these densities are defined as follows for a collection of point particles that give rise to continuous distributions of electric charge and current:

$$\rho(\vec{r}) = \sum_\alpha e_\alpha \delta(\vec{r} - \vec{q}_\alpha) \tag{2.7}$$

and

$$\vec{j}(\vec{r}) = \sum_\alpha e_\alpha \dot{\vec{q}}_\alpha \delta(\vec{r} - \vec{q}_\alpha). \tag{2.8}$$

In the last two relations, e_α is the charge of particle α positioned at $\vec{q}_\alpha$, and $\delta(\vec{r})$ is the Dirac delta function [3].

Auxiliary fields are absent from the microscopic Maxwell equations since all charges present in the system contribute to $\rho(\vec{r})$ and $\vec{j}(\vec{r})$. From the perspective of facilitating the quantisation of the electromagnetic field, it is beneficial to recast Maxwell's equations in terms of potentials rather than fields. This is done through the introduction of the scalar potential, $\phi(\vec{r}, t)$, and the vector potential, $\vec{a}(\vec{r}, t)$, on making use of the fact that a potential function is obtainable by integrating a field of force. Whence $\vec{b} = \mathrm{curl}\vec{a}$, and $-\nabla\phi = \vec{e} + \frac{\partial \vec{a}}{\partial t}$. Substituting these last two relations into the inhomogeneous Maxwell Eqs. (2.3) and (2.6) enable the potentials to be related to the sources. The potentials are themselves subject to transformation by the simultaneous addition of a gauge function, giving rise to a set of potentials ϕ and $\vec{a}$ which leave the fields $\vec{e}$ and $\vec{b}$ unchanged, and therefore Maxwell's equations invariant. A specific choice of gauge function is then said to fix the gauge.

A common choice, and one that will be adopted throughout this work is the Coulomb gauge, in which $\mathrm{div}\vec{a} = 0$. The potentials satisfy individual source dependent Maxwell's equations. ϕ obeys

$$\varepsilon_0 \nabla^2 \phi = -\rho, \tag{2.9}$$

which is immediately recognisable as Poisson's equation. Its solution represents the instantaneous Coulomb potential due to the charge density. Meanwhile $\vec{a}$ satisfies the wave equation

$$\left(\nabla^2 - \frac{1}{c^2} \frac{\partial^2}{\partial t^2} \right) \vec{a} = -\frac{1}{\varepsilon_0 c^2} \vec{j}^{\perp}, \tag{2.10}$$

where the transverse component of the current density $\vec{j}^{\perp}(\vec{r})$ appears in Eq. (2.10), and is a direct consequence of the transverse gauge condition and Helmholtz's theorem [4]. It may be extracted with the help of the transverse delta function dyadic

$$\delta_{ij}^{\perp}(\vec{r}) = \frac{1}{(2\pi)^3} \int (\delta_{ij} - \hat{k}_i \hat{k}_j) e^{i\vec{k}\cdot\vec{r}} d^3\vec{k} = (-\nabla^2 \delta_{ij} + \nabla_i \nabla_j) \frac{1}{4\pi r}, \tag{2.11}$$

with the longitudinal component given for completeness by

$$\delta_{ij}^{\parallel}(\vec{r}) = \frac{1}{(2\pi)^3} \int \hat{k}_i \hat{k}_j e^{i\vec{k}\cdot\vec{r}} d^3\vec{k} = -\nabla_i \nabla_j \frac{1}{4\pi r}, \tag{2.12}$$

the sum of the two yielding [6]

$$\delta_{ij}^{\perp}(\vec{r}) + \delta_{ij}^{\parallel}(\vec{r}) = \delta_{ij} \delta(\vec{r}). \tag{2.13}$$

In the last three formulae the Latin subscripts denote Cartesian tensor components. When indices repeat, a sum over each component is implied. Clearly evident from Eqs. (2.9) and (2.10) is the separation of the static and dynamic parts of the sources of the field in the Coulomb gauge, with $\vec{e}^{\perp} = -\partial \vec{a}^{\perp}/\partial t$, and $\vec{e}^{\parallel} = -\nabla \phi$. Incidentally, the vector potential is transverse in *all* gauges.

An interesting case occurs when both ρ and $\vec{j}$ vanish, corresponding to a free radiation field since there are no sources present. Solutions of Maxwell's equations then represent propagation of electromagnetic waves in vacuum. They are obtained by solving the wave equation

$$\left(\nabla^2 - \frac{1}{c^2} \frac{\partial^2}{\partial t^2} \right) \vec{e} = 0, \tag{2.14}$$

here written for the electric field, with a similar equation holding for $\vec{a}$ and $\vec{b}$. Plane wave solutions of the form

$$\vec{e}(\vec{r}, t) = \varepsilon \vec{e}^{(\lambda)}(\vec{k}) e^{i\vec{k}\cdot\vec{r} - i\omega t}, \tag{2.15}$$

follow straightforwardly, in which ε is the amplitude of the electric field, and $e^{(\lambda)}(\vec{k})$ is its complex unit electric polarisation vector for radiation propagating with wave vector $\vec{k}$ and index of polarisation λ, with circular frequency ω. Together $\vec{k}$ and λ describe the mode of the radiation field. Identical harmonic functional form solutions apply to the magnetic induction field and vector potential, the latter obtained from solution of Eq. (2.10) after setting the right-hand side equal to zero. The magnitude of $\vec{k}$ is $k = |\vec{k}| = \omega/c$. The unit electric and magnetic polarisation vectors, and direction of propagation, describe a right-handed triad, indicative of transverse wave propagation. Free electromagnetic radiation is in general elliptically polarised, but linear combinations of waves with appropriate choice of field strength and phase components readily produce linearly (plane) or circularly polarised light. To enumerate the allowed values of $\vec{k}$ to a countable infinity, radiation is confined to a cubic box of volume V, with the vector potential satisfying the periodic boundary condition that it have identical value on opposite sides of the box. The components of $\vec{k}$ are then restricted to $k_i = 2\pi n_i/l$, with n_i, $i = x, y, z$ taking on integer values, and l is the length of one side of the box.

In this section the laws underlying the classical mechanical behaviour and electromagnetic characteristics associated with charged particles have been summarized. The more interesting problem of interaction of microscopic forms of matter with the radiation field is examined next.

2.2 Lagrangian for a Charged Particle Coupled to Electromagnetic Radiation

Consider a particle α, with charge e_α, mass m_α and generalized coordinate $\vec{q}_\alpha$, and velocity $\dot{\vec{q}}_\alpha$, interacting with electromagnetic radiation described by scalar and vector potentials $\phi(\vec{r})$ and $\vec{a}(\vec{r})$. To facilitate construction of the QED Hamiltonian by means of the canonical quantization procedure, well known from particles only quantum mechanics [3], we begin by writing down the classical Lagrangian function for the particle, the electromagnetic field, and the interaction between the two as

$$L = L_{part} + L_{rad} + L_{int}. \tag{2.16}$$

Each of the three terms is given explicitly by:

$$L_{part} = \frac{1}{2}\sum_\alpha m_\alpha \dot{\vec{q}}_\alpha^2 - V(\vec{q}), \tag{2.17}$$

where $V(\vec{q})$ is the potential energy;

$$L_{rad} = \frac{1}{2}\varepsilon_0 \int \{\dot{\vec{a}}^2(\vec{r}) - c^2(\mathrm{curl}\vec{a}(\vec{r}))^2\}d^3\vec{r}, \tag{2.18}$$

and

$$L_{int} = \int \vec{j}^{\perp}(\vec{r}) \cdot \vec{a}(\vec{r}) d^3\vec{r}. \tag{2.19}$$

The choice of Lagrangian is justified if it leads to the correct equations of motion for the system under consideration. Since L in Eq. (2.16) is additive, it is instructive to consider each of the sub-systems individually before dealing with the total Lagrangian in the following section.

Assume for the moment that there is no radiation field. The last two terms of Eq. (2.16) consequently vanish. Substituting L_{part} from Eq. (2.17) into the Euler-Lagrange Eq. (2.2) produces for the equation of motion,

$$m_\alpha \frac{d^2 \vec{q}_\alpha}{dt^2} = -\frac{\partial V}{\partial \vec{q}_\alpha}, \tag{2.20}$$

which is immediately recognisable as Newton's Second Law of Motion, as is to be expected for non-relativistic kinematics. Proceeding with the canonical prescription in order to transition from classical to quantum mechanics, the next step involves the evaluation of the momentum canonically conjugate to the coordinate variable,

$$\vec{p}_\alpha = \frac{\partial L}{\partial \dot{\vec{q}}_\alpha}, \tag{2.21}$$

which for L_{part} above yields $\vec{p}_\alpha = m_\alpha \dot{\vec{q}}_\alpha$, for which kinetic and canonical momenta are equal.

Hamilton's principal function is then constructed via

$$H = \sum_\alpha \vec{p}_\alpha \dot{\vec{q}}_\alpha - L \tag{2.22}$$

after eliminating the velocity in favour of the momentum. Hamilton's canonical equations follow on taking the total derivative of H in Eq. (2.22), giving

$$\dot{\vec{q}}_\alpha = \frac{\partial H}{\partial \vec{p}_\alpha}, \tag{2.23}$$

$$\dot{\vec{p}}_\alpha = -\frac{\partial H}{\partial \vec{q}_\alpha}, \tag{2.24}$$

and

$$\frac{\partial H}{\partial t} = -\frac{\partial L}{\partial t}, \tag{2.25}$$

if L is explicitly time-dependent. Continuing with this particles only scenario, inserting the velocity and L_{part} in Eq. (2.22) results in the particle Hamiltonian

$$H_{part} = \sum_\alpha \frac{1}{2m_\alpha} \vec{p}_\alpha^2 + V(\vec{q}), \tag{2.26}$$

which represents the total classical energy of a conservative system, and is a sum of kinetic and potential energy contributions.

Let us now assume that there are no sources of charge and current. Only L_{rad} of Eq. (2.16) therefore remains, corresponding to the free radiation field. Choosing the vector potential to be the analogue of the "coordinate" variable, and its time derivative to be the "velocity" variable, the canonical formalism valid for particles may be applied to the electromagnetic field. Because the field is continuous, neighbouring points in space are related via the spatial gradient as well as by the displacement between them. Accounting for this fact results in the Euler-Lagrange Eq. (2.2) being modified by an additional term, which for the ith component reads in total as

$$\frac{\partial}{\partial t} \left(\frac{\partial L^D}{\partial \dot{a}_i} \right) + \frac{\partial}{\partial x_j} \frac{\partial L^D}{\partial (\partial a_i / \partial x_j)} - \frac{\partial L^D}{\partial a_i} = 0. \tag{2.27}$$

L^D is a Lagrangian density. Its integral over all space yields L. With L_{rad}^D from Eq. (2.18) given by $\frac{1}{2}\varepsilon_0[\dot{\vec{a}}^2 - c^2(\text{curl}\vec{a})^2]$, application of Eq. (2.27) gives rise to the source free wave equation

$$\left(\nabla^2 - \frac{1}{c^2} \frac{\partial^2}{\partial t^2} \right) a_i = 0, \tag{2.28}$$

validating the choice of radiation field Lagrangian Eq. (2.18).

Analogously to Eq. (2.21), the momentum canonically conjugate to the field coordinate is defined as

$$\vec{\Pi}(\vec{r}) = \frac{\partial L^D}{\partial \dot{\vec{a}}(\vec{r})}. \tag{2.29}$$

From L_{rad}^D, $\vec{\Pi}(\vec{r})$ is explicitly found in this case to be

$$\vec{\Pi}(\vec{r}) = \varepsilon_0 \dot{\vec{a}}(\vec{r}) = -\varepsilon_0 \vec{e}^\perp(\vec{r}). \tag{2.30}$$

Proceeding with the canonical formulation, the classical Hamiltonian for the free radiation field is then calculated from

$$H_{rad} = \int (\vec{\Pi}(\vec{r}) \cdot \dot{\vec{a}}(\vec{r}) - L_{rad}^D) d^3\vec{r}, \tag{2.31}$$

with the integrand corresponding to the Hamiltonian density. Eliminating $\dot{\vec{a}}$ in terms of $\vec{\Pi}$ from Eq. (2.30), H_{rad} when expressed in terms of canonical variables is written as

$$H_{rad} = \frac{1}{2} \int \{[\vec{\Pi}^2(\vec{r})/\varepsilon_0] + \varepsilon_0 c^2 [\text{curl}\vec{a}(\vec{r})]^2\}d^3\vec{r}. \tag{2.32}$$

On making use of the definition of the vector potential, and the right-hand most form of Eq. (2.30), H_{rad} can be written more transparently as functions of the electric and magnetic induction fields, as in

$$H_{rad} = \frac{\varepsilon_0}{2} \int \{\vec{e}^{\perp 2}(\vec{r}) + c^2 \vec{b}^2(\vec{r})\}d^3\vec{r}. \tag{2.33}$$

Either of the last two integrands provides an expression for the electromagnetic energy density.

Interestingly, it was recognised by Born, Heisenberg and Jordan [7] that the quantum mechanical version of H_{rad} is equivalent to the Hamiltonian of a mechanically vibrating system, as demonstrated by Jeans' theorem in the classical regime [8]. By defining two real variables

$$q_{\vec{k}}^{(\lambda)} = (\varepsilon_0 V)^{1/2}(a_{\vec{k}}^{(\lambda)} + \bar{a}_{\vec{k}}^{(\lambda)}) \tag{2.34}$$

$$p_{\vec{k}}^{(\lambda)} = -i\omega(\varepsilon_0 V)^{1/2}(a_{\vec{k}}^{(\lambda)} - \bar{a}_{\vec{k}}^{(\lambda)}), \tag{2.35}$$

where $a_{\vec{k}}^{(\lambda)}$ are complex Fourier mode components [4] of the vector potential, with the overbar denoting the complex conjugate, and V is the quantisation volume, H_{rad} may be re-expressed as a sum of simple harmonic oscillator Hamiltonians, one for each mode $(\vec{k}, \lambda)$

$$H_{rad} = \sum_{\vec{k},\lambda} \frac{1}{2} \{p_{\vec{k}}^{(\lambda)2} + \omega^2 q_{\vec{k}}^{(\lambda)2}\} = \sum_{\vec{k},\lambda} H_{\vec{k},\lambda}. \tag{2.36}$$

The new variables (2.34) and (2.35) are canonically conjugate, and result in the correct Hamilton's equations of motion being obtained on using (2.23) and (2.24).

Having established that L_{part} and L_{rad} contained in the total Lagrangian Eq. (2.16) each correctly describe the dynamics in the absence of a radiation field, and when there are no sources present, respectively, in the next section we examine the coupled matter-electromagnetic field system, and see how the equations of motion are modified due to interaction. The system is then quantised and a QED Hamiltonian operator is finally obtained.

2.3 Minimal-Coupling QED Hamiltonian

We now verify that the total Lagrangian for the interacting system, Eq. (2.16), leads to the correct equations of motion, appropriately changed to account for the inclusion of L_{int}. Application of Eq. (2.2) gives rise to [9]

$$m_\alpha \frac{d^2 q_{i(\alpha)}}{dt^2} = -\frac{\partial V}{\partial q_{i(\alpha)}} + e_\alpha e_i^\perp(\vec{q}_\alpha) + e_\alpha \left(\frac{d\vec{q}_\alpha}{dt} \times \vec{b}(\vec{q}_\alpha)\right)_i, \qquad (2.37)$$

instead of Eq. (2.20). Newton's equations of motion are now modified by the addition of Lorentz force law terms describing the coupling of the charged particle to the electromagnetic field. Application of Eq. (2.27) to the full Lagrangian (2.16) changes Eq. (2.28) to Eq. (2.10), the expected wave equation satisfied by the vector potential in the presence of sources.

The classical Hamiltonian function for the coupled system may be obtained by following the canonical scheme implemented in the previous section. The particle momentum is no longer equal to its kinetic momentum, but changes to

$$\vec{p}_\alpha = m_\alpha \frac{d\vec{q}_\alpha}{dt} + e_\alpha \vec{a}(\vec{q}_\alpha). \qquad (2.38)$$

$\vec{\Pi}(\vec{r})$, however, remains identical to Eq. (2.30). H is then calculated from

$$H = \sum_\alpha \vec{p}_\alpha \cdot \dot{\vec{q}}_\alpha + \int \vec{\Pi}(\vec{r}) \cdot \dot{\vec{a}}(\vec{r}) d^3\vec{r} - L, \qquad (2.39)$$

which for a many-particle system is found to be [9, 10]

$$H = \sum_\alpha \frac{1}{2m_\alpha} \{\vec{p}_\alpha - e_\alpha \vec{a}(\vec{q}_\alpha)\}^2 + V(\vec{q}) + \frac{1}{2} \int \{\frac{\vec{\Pi}^2}{\varepsilon_0} + \varepsilon_0 c^2 (\text{curl}\vec{a})^2\} d^3\vec{r}. \qquad (2.40)$$

Equation (2.40) is known as the minimal-coupling Hamiltonian on account of the minimum action principle being applied to its construction. Coupling of radiation with matter simply amounts to replacing the particle momentum by $\vec{p}_\alpha - e_\alpha \vec{a}(\vec{q}_\alpha)$.

At this point in the development it is convenient to collect the charged particles α and form atoms and molecules ξ. Furthermore, the nuclei are henceforth taken to be stationary, with the positions and momenta of the electrons only being considered. As a result, the Hamiltonian Eq. (2.40) may be partitioned as [9, 10]

$$H = \sum_\xi H_{part}(\xi) + H_{rad} + \sum_{\substack{\xi,\xi' \\ \xi<\xi'}} H_{int}(\xi, \xi'), \qquad (2.41)$$

into a sum of particle, radiation field, and interaction contributions. Decomposing the electrostatic energy into a sum of single- and two-particle terms,

$$V = V(\xi) + V(\xi, \xi'), \tag{2.42}$$

enables the first term of Eq. (2.41) to be written as

$$H_{part}(\xi) = \sum_{\alpha} \frac{1}{2m_{\alpha}} \vec{p}_{\alpha}^{2}(\xi) + V(\xi), \tag{2.43}$$

and which differs slightly from Eq. (2.26) in that $V(\xi)$ is now interpreted as the intra-molecular potential energy. Hence Eq. (2.43) corresponds to the familiar molecular Hamiltonian of non-relativistic quantum mechanics in the Born-Oppenheimer approximation [11]. The second term of Eq. (2.41) is given by the third term of Eq. (2.40) and is seen to be identical to H_{rad} calculated for the free radiation field Eq. (2.32). The remaining terms of Eq. (2.40), along with the second term of Eq. (2.42) constitute the interaction Hamiltonian,

$$H_{int}(\xi, \xi') = -\sum_{\alpha} \frac{e_{\alpha}}{m_{\alpha}} \vec{p}_{\alpha}(\xi) \cdot \vec{a}(\vec{q}_{\alpha}(\xi)) + \sum_{\alpha} \frac{e_{\alpha}^{2}}{2m_{\alpha}} \vec{a}^{2}(\vec{q}_{\alpha}(\xi)) + V(\xi, \xi'), \tag{2.44}$$

where the third term of Eq. (2.44) is the instantaneous inter-particle Coulomb potential. Even though $V(\xi, \xi')$ appears explicitly, a fully retarded result is obtained on using Eq. (2.44) on account of exact cancellation of static terms with those arising from the vector potential, which contains non-retarded contributions in the Coulomb gauge. The first two terms of the interaction Hamiltonian are linear and quadratic in the vector potential evaluated at the position of electron α in atom or molecule ξ.

Quantisation of the classical minimal-coupling Hamiltonian (2.41) then follows by promoting the classical dynamical variables, for both particles and radiation field, namely the respective coordinates $\vec{q}_{\alpha}(\xi)$ and $\vec{a}(\vec{r})$, and momenta $\vec{p}_{\alpha}(\xi)$ and $\vec{\Pi}(\vec{r})$, to quantum mechanical operators subject to the canonical equal time commutation relations

$$[q_{i(\alpha)}(\xi), p_{j(\beta)}(\xi')] = i\hbar \delta_{ij} \delta_{\alpha\beta} \delta_{\xi\xi'}, \tag{2.45}$$

and

$$[a_{i}(\vec{r}), \Pi_{j}(\vec{r}')] = i\hbar \delta_{ij}^{\perp}(\vec{r} - \vec{r}'). \tag{2.46}$$

Recalling that the equations describing the electromagnetic field are formally equivalent to that of an oscillating mechanical system, quantisation of the radiation field corresponds to quantisation of the simple harmonic oscillator Hamiltonian Eq. (2.36). A powerful and elegant method of obtaining eigenvalue and eigenfunction solutions to this problem is through the techniques of second quantisation via the introduction of lowering and raising operators [12]

$$a = \frac{1}{\sqrt{2}} \left(\sqrt{\frac{m\omega}{\hbar}} q + i \sqrt{\frac{1}{m\omega\hbar}} p \right), \qquad (2.47)$$

and

$$a^\dagger = \frac{1}{\sqrt{2}} \left(\sqrt{\frac{m\omega}{\hbar}} q - i \sqrt{\frac{1}{m\omega\hbar}} p \right), \qquad (2.48)$$

which are real and mutually adjoint, but are not symmetric and therefore non-Hermitian. Thus the radiation field Hamiltonian Eq. (2.36) can be expressed in terms of $(\vec{k}, \lambda)-$ mode annihilation and creation operators $a^{(\lambda)}(\vec{k})$ and $a^{\dagger(\lambda)}(\vec{k})$ as

$$H_{rad} = \sum_{\vec{k},\lambda} \{ a^{\dagger(\lambda)}(\vec{k}) a^{(\lambda)}(\vec{k}) + \frac{1}{2} \} \hbar\omega, \qquad (2.49)$$

subject to the commutator

$$[a^{(\lambda)}(\vec{k}), a^{\dagger(\lambda')}(\vec{k}')] = \delta_{\lambda\lambda'} \delta(\vec{k} - \vec{k}'), \qquad (2.50)$$

with all other boson operator combinations commuting.

Identification of the operator combination $a^{\dagger(\lambda)}(\vec{k}) a^{(\lambda)}(\vec{k})$ as the number operator $n(\vec{k}, \lambda)$, allows the eigenvalue spectrum of the quantised electromagnetic field to be written down immediately as $(n + \frac{1}{2})\hbar\omega$, $n = 0, 1, 2, \ldots$. The excitation quanta therefore correspond to the number of photons in the radiation field, which is characteristic of the occupation number representation adopted. The state of the electromagnetic field is specified by the ket $|n(\vec{k}, \lambda)\rangle$. It is usual to suppress states with zero photons. The bosonic operators $a^{(\lambda)}(\vec{k})$ and $a^{\dagger(\lambda)}(\vec{k})$, respectively decrease or increase the number of photons in the radiation field by unity according to the operator relations [3, 6]

$$\begin{aligned} a^{(\lambda)}(\vec{k})|n(\vec{k}, \lambda)\rangle &= 0, & n = 0, \\ &= n^{1/2}|(n-1)(\vec{k}, \lambda)\rangle, & n = 1, 2, \ldots \end{aligned} \qquad (2.51)$$

and

$$a^{\dagger(\lambda)}(\vec{k})|n(\vec{k},\lambda)\rangle = (n+1)^{1/2}|(n+1)(\vec{k},\lambda)\rangle, \quad n = 0,1,2,\dots, \quad (2.52)$$

along with the number operator

$$a^{\dagger(\lambda)}(\vec{k})a^{(\lambda)}(\vec{k})|n(\vec{k},\lambda)\rangle = n|n(\vec{k},\lambda)\rangle, \quad n = 0,1,2,\dots. \quad (2.53)$$

It is interesting to note that the state of the radiation field in which there are no photons represents its ground state, corresponding to the electromagnetic vacuum, for which $n = 0$ [13]. An ever present $\frac{1}{2}\hbar\omega$ of zero-point energy is associated per mode of the field, resulting in an infinite ground state field energy. Nonetheless, measurable effects ensue from vacuum fluctuations, with the van der Waals dispersion force perhaps being one of the most important. Others include spontaneous emission (since $a^{\dagger(\lambda)}(\vec{k})$ can act on the vacuum state to create a photon), and the Lamb shift [1].

A Fourier mode expansion [4] of the vector potential as a function of the creation and destruction operators in the Schrödinger picture takes the form

$$\vec{a}(\vec{r}) = \sum_{\vec{k},\lambda} \left(\frac{\hbar}{2\varepsilon_0 ckV}\right)^{1/2} [\vec{e}^{(\lambda)}(\vec{k})a^{(\lambda)}(\vec{k})e^{i\vec{k}\cdot\vec{r}} + \vec{e}^{(\lambda)}(\vec{k})a^{\dagger(\lambda)}(\vec{k})e^{-i\vec{k}\cdot\vec{r}}], \quad (2.54)$$

where $\vec{e}^{(\lambda)}(\vec{k})$ is a complex unit electric polarisation vector for a $(\vec{k},\lambda)-$ mode photon, and V is the box quantisation volume. Similar expressions for $\vec{e}^{\perp}(\vec{r})$, $\vec{b}(\vec{r})$ and $\vec{\Pi}(\vec{r})$ follow from their definitions in terms of $\vec{a}(\vec{r})$ given earlier, namely $\vec{e}^{\perp}(\vec{r}) = -\dot{\vec{a}}(\vec{r}) = -\varepsilon_0^{-1}\vec{\Pi}(\vec{r})$, and $\vec{b}(\vec{r}) = \text{curl}\vec{a}(\vec{r})$. They are

$$\vec{e}^{\perp}(\vec{r}) = i\sum_{\vec{k},\lambda} \left(\frac{\hbar ck}{2\varepsilon_0 V}\right)^{1/2} [\vec{e}^{(\lambda)}(\vec{k})a^{(\lambda)}(\vec{k})e^{i\vec{k}\cdot\vec{r}} - \vec{e}^{(\lambda)}(\vec{k})a^{\dagger(\lambda)}(\vec{k})e^{-i\vec{k}\cdot\vec{r}}], \quad (2.55)$$

$$\vec{b}(\vec{r}) = i\sum_{\vec{k},\lambda} \left(\frac{\hbar k}{2\varepsilon_0 cV}\right)^{1/2} [\vec{b}^{(\lambda)}(\vec{k})a^{(\lambda)}(\vec{k})e^{i\vec{k}\cdot\vec{r}} - \vec{b}^{(\lambda)}(\vec{k})a^{\dagger(\lambda)}(\vec{k})e^{-i\vec{k}\cdot\vec{r}}], \quad (2.56)$$

where the unit magnetic polarisation vector is $\vec{b}^{(\lambda)}(\vec{k}) = \hat{k} \times \vec{e}^{(\lambda)}(\vec{k})$, and

$$\vec{\Pi}(\vec{r}) = -i\sum_{\vec{k},\lambda} \left(\frac{\hbar ck\varepsilon_0}{2V}\right)^{1/2} [\vec{e}^{(\lambda)}(\vec{k})a^{(\lambda)}(\vec{k})e^{i\vec{k}\cdot\vec{r}} - \vec{e}^{(\lambda)}(\vec{k})a^{\dagger(\lambda)}(\vec{k})e^{-i\vec{k}\cdot\vec{r}}], \quad (2.57)$$

the first factor after each sum in the mode expansion ensures that each field is correctly normalised to reproduce the energy of the electromagnetic field.

2.4 Multipolar-Coupling QED Hamiltonian

While the minimal-coupling Hamiltonian Eq. (2.41) may be employed to compute light-matter and inter-particle interactions, the form of coupling Hamiltonian Eq. (2.44) isn't the most advantageous to work with from the point of view of treating chemical systems. Inspection of $H_{int}(\xi, \xi')$ reveals the presence of the particle momentum operator in the first term, the vector potential operator and its square in the first and second terms, respectively, and the instantaneous two-particle coupling $V(\xi, \xi')$ given by the last contribution of Eq. (2.44). A superior alternative QED Hamiltonian is provided by the multipolar counterpart. Here atoms and molecules couple directly to the causal Maxwell field operators through their molecular multipole moment distributions, and all instantaneous couplings have been eliminated. Use of either Hamiltonian leads to results that are properly retarded. In the minimal-coupling scheme this occurs through explicit cancellation of static contributions. The multipolar version may be obtained from the minimal-coupling form by applying a quantum canonical transformation on H_{min}, Eq. (2.40), using a generating function S that is independent of time. Although the new Hamiltonian differs in functional form relative to the old one, identical eigenspectra result with the use of either Hamiltonian since the transformation is unitary. It is of the form

$$e^{iS} H_{min} e^{-iS} = H_{mult}. \tag{2.58}$$

Hamiltonians related in this way are said to be equivalent [14]. An intrinsic feature of quantum canonical transformations is that they leave the commutator between canonically conjugate dynamical variables invariant, for instance

$$[q, p] = i\hbar, \tag{2.59}$$

and they leave the Heisenberg operator equations of motion unchanged, the latter being the quantum versions of Hamilton's canonical equations,

$$i\hbar \dot{q} = [q, H], \tag{2.60}$$

and

$$i\hbar \dot{p} = [p, H]. \tag{2.61}$$

It is easily verified that transformation (2.58) guarantees that these properties are satisfied [9]. Hence a quantum canonical transformation in essence amounts to transforming the original canonically conjugate dynamical variables of the system and expressing the original Hamiltonian in terms of the newly transformed quantities. It is therefore the quantum mechanical analogue of a contact transformation in classical mechanics [2, 3].

The specific form of the generator S that enables transformation of minimal-coupling variables to those in the multipolar formalism is [15]

$$S = \frac{1}{\hbar} \int \vec{p}^{\perp}(\vec{r}) \cdot \vec{a}(\vec{r}) d^3 \vec{r}. \tag{2.62}$$

In Eq. (2.62), $\vec{p}^{\perp}(\vec{r})$ is the transverse component of the electric polarisation field for a molecular assembly,

$$\vec{p}(\vec{r}) = \sum_{\xi} \vec{p}(\xi; \vec{r}), \tag{2.63}$$

where a closed form expression for the electronic part of $\vec{p}(\xi; \vec{r})$, written in terms of a parametric integral, is [16]

$$\vec{p}(\xi; \vec{r}) = -e \sum_{\alpha} (\vec{q}_{\alpha}(\xi) - \vec{R}_{\xi}) \int_0^1 \delta(\vec{r} - \vec{R}_{\xi} - \lambda(\vec{q}_{\alpha}(\xi) - \vec{R}_{\xi})) d\lambda, \tag{2.64}$$

where $\vec{R}_{\xi}$ is the position vector of the centre of species ξ. Because the generator is a function only of the particle and field coordinates, $\vec{q}_{\alpha}(\xi)$ and $\vec{a}(\vec{r})$ remain unchanged by the transformation, with only the canonically conjugate momenta being transformed.

Employing the Baker-Hausdorff identity for two non-commuting operators A and B,

$$e^A B e^{-A} = B + [A, B] + \frac{1}{2!}[A, [A, B]] + \frac{1}{3!}[A, [A, [A, B]]] + \cdots, \tag{2.65}$$

it can be shown [9] that the particle momentum transforms as

$$\vec{p}_{\alpha}^{mult}(\xi) = e^{iS} \vec{p}_{\alpha}^{min}(\xi) e^{-iS} = \vec{p}_{\alpha}^{min}(\xi) + i[S, \vec{p}_{\alpha}^{min}(\xi)] + \cdots$$
$$= \vec{p}_{\alpha}^{min}(\xi) + e\vec{a}(\vec{q}_{\alpha}(\xi)) - \int \vec{n}_{\alpha}(\xi; \vec{r}) \times \vec{b}(\vec{r}) d^3 \vec{r}, \tag{2.66}$$

on inserting the generator (2.62). In Eq. (2.66), the vector field $\vec{n}_{\alpha}(\xi; \vec{r})$ is defined as

$$\vec{n}_{\alpha}(\xi; \vec{r}) = -e(\vec{q}_{\alpha}(\xi) - \vec{R}_{\xi}) \int_0^1 \lambda \delta(\vec{r} - \vec{R}_{\xi} - \lambda(\vec{q}_{\alpha}(\xi) - \vec{R}_{\xi})) d\lambda, \tag{2.67}$$

with

$$\vec{n}(\vec{r}) = \sum_{\xi,\alpha} \vec{n}_\alpha(\xi;\vec{r}). \tag{2.68}$$

Carrying out the transformation on the field momentum, we find in similar fashion

$$\vec{\Pi}^{mult}(\vec{r}) = e^{iS}\vec{\Pi}^{min}(\vec{r})e^{-iS} = \vec{\Pi}^{min}(\vec{r}) + i[S,\vec{\Pi}^{min}(\vec{r})] + \cdots$$

$$= \Pi^{min}(\vec{r}) + \frac{i}{\hbar}\Big[\int \vec{p}^\perp(\vec{r}') \cdot \vec{a}(\vec{r}')d^3\vec{r}', \vec{\Pi}^{min}(\vec{r})\Big] + \cdots. \tag{2.69}$$

On using the field commutation relation (2.46), it is seen that S commutes with the minimal-coupling field momentum so that all subsequent higher-order nested commutators not explicitly written in Eq. (2.69), but present in Eq. (2.65) vanish, leaving the canonically conjugate field momentum in multipolar framework as

$$\vec{\Pi}^{mult}(\vec{r}) = \vec{\Pi}^{min}(\vec{r}) - \vec{p}^\perp(\vec{r}). \tag{2.70}$$

Recalling from Eq. (2.30) that $\vec{\Pi}^{min}(\vec{r}) = -\varepsilon_0\vec{e}^\perp(\vec{r})$, it is seen that

$$\vec{\Pi}^{mult}(\vec{r}) = -\vec{d}^\perp(\vec{r}), \tag{2.71}$$

where $\vec{d}^\perp(\vec{r})$ is the transverse component of the electric displacement field $\vec{d}(\vec{r})$ defined by

$$\vec{d}(\vec{r}) = \varepsilon_0\vec{e}(\vec{r}) + \vec{p}(\vec{r}). \tag{2.72}$$

Hence in the multipolar formalism, the canonically conjugate field momentum is no longer proportional to the transverse electric field, but is instead equal to the negative of the displacement field. A mode expansion for this last field quantity is given by

$$\vec{d}^\perp(\vec{r}) = i\sum_{\vec{k},\lambda}\left(\frac{\hbar ck\varepsilon_0}{2V}\right)^{1/2}[\vec{e}^{(\lambda)}(\vec{k})a^{(\lambda)}(\vec{k})e^{i\vec{k}\cdot\vec{r}} - \vec{e}^{(\lambda)}(\vec{k})a^{\dagger(\lambda)}(\vec{k})e^{-i\vec{k}\cdot\vec{r}}]. \tag{2.73}$$

The multipolar form of Hamiltonian follows on substituting for $\vec{p}_\alpha^{mult}(\xi)$, Eq. (2.66) and $\vec{\Pi}^{mult}(\vec{r})$, Eq. (2.70) into the minimal-coupling Hamiltonian Eq. (2.40),

$$H^{mult} = \sum_{\xi} \frac{1}{2m} \sum_{\alpha} \{\vec{p}_{\alpha}(\xi) + \int \vec{n}_{\alpha}(\xi;\vec{r}) \times \vec{b}(\vec{r}) d^3\vec{r}\}^2 + \sum_{\xi} V(\xi)$$

$$+ \frac{1}{2\varepsilon_0} \int \{[\vec{\Pi}(\vec{r}) + \vec{p}^{\perp}(\vec{r})]^2 + \varepsilon_0^2 c^2 (\text{curl} \vec{a}(\vec{r}))^2\} d^3\vec{r} + \sum_{\substack{\xi,\xi' \\ \xi > \xi'}} V(\xi,\xi'), \qquad (2.74)$$

on employing decomposition (2.42). As done for its minimal-coupling counterpart, H^{mult} may be partitioned into particle, radiation field, and interaction Hamiltonian terms. Thus

$$H^{mult} = H^{mult}_{part} + H^{mult}_{rad} + H^{mult}_{int} + \frac{1}{2\varepsilon_0} \int \sum_{\xi} |\vec{p}^{\perp}(\xi;\vec{r})|^2 d^3\vec{r}, \qquad (2.75)$$

with

$$H^{mult}_{part} = \sum_{\xi} \{\frac{1}{2m} \sum_{\alpha} \vec{p}_{\alpha}^2(\xi) + V(\xi)\}, \qquad (2.76)$$

and

$$H^{mult}_{rad} = \frac{1}{2\varepsilon_0} \int \{\vec{d}^{\perp 2}(\vec{r}) + \varepsilon_0^2 c^2 \vec{b}^2(\vec{r})\} d^3\vec{r}, \qquad (2.77)$$

when written explicitly in terms of Maxwell fields, and

$$H^{mult}_{int} = -\varepsilon_0^{-1} \int \vec{p}(\vec{r}) \cdot \vec{d}^{\perp}(\vec{r}) d^3\vec{r}$$

$$- \int \vec{m}(\vec{r}) \cdot \vec{b}(\vec{r}) d^3\vec{r} + \frac{1}{2} \int O_{ij}(\vec{r},\vec{r}') b_i(\vec{r}) b_j(\vec{r}') d^3\vec{r} d^3\vec{r}'. \qquad (2.78)$$

$\vec{p}(\vec{r})$ appearing in the first term of H^{mult}_{int} is the electric polarisation field (2.63). Two new fields feature in the remaining contributions. One is the magnetisation field,

$$\vec{m}(\vec{r}) = \frac{1}{2m} \sum_{\xi,\alpha} [\vec{n}_{\alpha}(\xi;\vec{r}) \times \vec{p}_{\alpha}(\xi) - \vec{p}_{\alpha}(\xi) \times \vec{n}_{\alpha}(\xi;\vec{r})], \qquad (2.79)$$

and the other is the diamagnetisation field

$$O_{ij}(\vec{r},\vec{r}') = \sum_{\xi} \sum_{\xi'} \frac{1}{m} \varepsilon_{ikl} \varepsilon_{jml} n_k(\xi;\vec{r}) n_m(\xi';\vec{r}'), \qquad (2.80)$$

where the α-dependence of the n_i tensors is implicit. A remarkable aspect of the multipolar Hamiltonian relative to its minimal-coupling precursor is the disappearance of the static inter-particle coupling term $V(\xi, \xi')$. On separating the square of the transverse polarisation contribution from (2.74) into a sum of intra- and inter-molecular parts, it is found that the inter-molecular contribution exactly cancels $V(\xi, \xi')$. What remains is a one-centre transverse electric polarisation field squared contribution, the last term of Eq. (2.75). Because it is independent of the radiation field, it may be neglected when considering effects that produce a change in the state of the electromagnetic field. When evaluating corrections to the self-energy, however, this contribution must be retained. Another feature of Eq. (2.75) is the explicit presence of the transverse electric displacement field. This is a direct consequence of the transformed field momentum Eq. (2.70). Furthermore, atoms and molecules couple directly to the electric displacement and magnetic induction fields through electric polarisation, magnetisation and dia-magnetisation distributions. Because the Maxwell fields are strictly causal, inter-actions between centres of charge and current are properly retarded, with electromagnetic signals propagating at the correct speed, namely that of light, c. There are no static coupling terms.

The transformation of the minimal-coupling Hamiltonian to H^{mult} via Eq. (2.58) and the generator (2.62) is known as the Power-Zienau-Woolley transformation [9, 10, 14–21]. An alternative method of arriving at H^{mult} is to first transform the minimal-coupling Lagrangian, Eq. (2.16). This may be accomplished by the addition to L of the time derivative of a function of the coordinates and the time only. The effect of such a modification is to generate an equivalent Lagrangian, in the sense that the Euler-Lagrange equations of motion (2.2) remain identical in form [2]. Again the coordinate variable remains invariant, but the new canonically conjugate momentum is changed to $\vec{p} + \frac{\partial f(\vec{q},t)}{\partial \vec{q}}$, where f is the transformation function.

The general connection between f and the generator S that yields equivalent Hamiltonians is easily found. Applying the first line of Eq. (2.66), the new momentum obtained via canonical transformation is $\vec{p} - \hbar \frac{\partial S}{\partial \vec{q}}$, so that $f = -\hbar S$. Hence the multipolar Hamiltonian Eq. (2.75) will result when $-\frac{d}{dt} \int \vec{p}^{\perp}(\vec{r}) \cdot \vec{a}(\vec{r}) d^3\vec{r}$ is added to L_{min}, Eq. (2.16) to give L_{mult} [14, 22, 23]. This addition has the desired effect of removing coupling via the transverse current. On account of

$$\vec{j}(\vec{r}) = \frac{d\vec{p}(\vec{r})}{dt} + \mathrm{curl}\vec{m}(\vec{r}), \qquad (2.81)$$

interaction now occurs through the polarisation and magnetisation fields instead. The newly transformed Lagrangian, L_{mult}, leads to the correct equations of motion. In the case of particles, this is the Newton-Lorentz force law Eq. (2.37). For the radiation field, the resulting equations are known as the atomic field equations. They are intermediate between the microscopic Maxwell-Lorentz Eqs. (2.3)–(2.6), and the macroscopic Maxwell equations. The source free Maxwell Eqs. (2.4) and (2.5) are trivially satisfied by the form of the electromagnetic potentials in the Coulomb gauge. Meanwhile Eq. (2.3) becomes

$$\mathrm{div}\vec{d}(\vec{r}) = \rho^{true}(\vec{r}), \tag{2.82}$$

where $\vec{d}(\vec{r})$ was defined by Eq. (2.72), showing that the true charges are solely responsible for the electric displacement field. The second source dependent Maxwell-Lorentz Eq. (2.6) becomes

$$\mathrm{curl}\vec{h}(\vec{r}) = \frac{\partial \vec{d}^{\perp}(\vec{r})}{\partial t}, \tag{2.83}$$

on defining the auxiliary magnetic field

$$\vec{h}(\vec{r}) = \varepsilon_0 c^2 \vec{b}(\vec{r}) - \vec{m}(\vec{r}), \tag{2.84}$$

after taking the transverse component of relation (2.81), thereby ensuring the implicit presence of the current density. Hence the sources are represented by electric polarisation and magnetisation distributions, which are in turn the origins of the microscopic fields $\vec{d}(\vec{r})$ and $\vec{h}(\vec{r})$.

By expanding $\vec{p}(\vec{r})$ and $\vec{m}(\vec{r})$ about $\vec{R}_{\xi}$ in a Taylor series [5], and retaining the first few terms in the series, electric dipole and quadrupole polarisation distributions,

$$\vec{p}(\xi;\vec{r}) = \sum_{\alpha} e_{\alpha}(\vec{q}_{\alpha}(\xi) - \vec{R}_{\xi})\{1 - \frac{1}{2!}(\vec{q}_{\alpha}(\xi) - \vec{R}_{\xi}) \cdot \nabla + \cdots\}\delta(\vec{r} - \vec{R}_{\xi}), \tag{2.85}$$

and the magnetic dipole contribution to the magnetisation field,

$$\vec{m}(\xi;\vec{r}) = \sum_{\alpha} \frac{e_{\alpha}}{m_{\alpha}}\{(\vec{q}_{\alpha}(\xi) - \vec{R}_{\xi}) \times \vec{p}_{\alpha}(\xi)\}\{\frac{1}{2!} - \cdots\}\delta(\vec{r} - \vec{R}_{\xi}), \tag{2.86}$$

ensue, and similarly for the diamagnetisation distribution, $O_{ij}(\vec{r},\vec{r}')$. Integrating over all space yields for the interaction Hamiltonian Eq. (2.78), the multipole expanded form

$$\begin{aligned} H_{int}^{mult}(\xi) = &-\varepsilon_0^{-1}\vec{\mu}(\xi) \cdot \vec{d}^{\perp}(\vec{R}_{\xi}) - \varepsilon_0^{-1}Q_{ij}(\xi)\nabla_j d_i^{\perp}(\vec{R}_{\xi}) + \cdots \\ &-\vec{m}(\xi) \cdot \vec{b}(\vec{R}_{\xi}) + \cdots \\ &+ \frac{e^2}{8m}\sum_{\alpha}\{(\vec{q}_{\alpha}(\xi) - \vec{R}_{\xi}) \times \vec{b}(\vec{R}_{\xi})\}^2 + \cdots \end{aligned}, \tag{2.87}$$

where the electric dipole moment operator,

$$\vec{\mu}(\xi) = -\frac{e}{1!}\sum_{\alpha}(\vec{q}_{\alpha}(\xi) - \vec{R}_{\xi}), \tag{2.88}$$

the electric quadrupole moment is

$$Q_{ij}(\xi) = -\frac{e}{2!}\sum_\alpha (\vec{q}_\alpha(\xi) - \vec{R}_\xi)_i (\vec{q}_\alpha(\xi) - \vec{R}_\xi)_j, \qquad (2.89)$$

the magnetic dipole moment is

$$\vec{m}(\xi) = \frac{1}{2!}\sum_\alpha \frac{e_\alpha}{m_\alpha}\{(\vec{q}_\alpha(\xi) - \vec{R}_\xi) \times \vec{p}_\alpha(\xi)\}, \qquad (2.90)$$

and the last term written explicitly in Eq. (2.87) is the leading order diamagnetic contribution.

If the dimensions of ξ are very small relative to the wavelength of light, it is sufficient to keep only the first coupling term of Eq. (2.87), in what is known as the electric dipole approximation. This is further justified on account of $\vec{m}(\xi)$ and $\vec{\overset{\leftrightarrow}{Q}}(\xi)$ typically being of the order of the fine structure constant smaller than $\vec{\mu}(\xi)$, these higher-order multipolar terms providing small corrections to the electric dipole interaction term. For improved accuracy, and when treating optically active species, however, it is necessary to include contributions from higher multipole moments.

2.5   Perturbative Solution to the QED Hamiltonian

If the coupling between electromagnetic radiation and matter is taken to be weak compared to the strength of intra-atomic or molecular Coulomb fields, as is frequently the case if the magnitude of the electric field strength of the radiation is of the order of 10^6 Vcm^{-1} or less, eigenvalue and eigenfunction solutions to the QED Hamiltonian operator may be obtained perturbatively. The total Hamiltonian is separated into an unperturbed part, H_0, comprising the sum of H_{part} and H_{rad}, and the perturbation operator given by H_{int}, namely

$$H = H_0 + H_{int}, \qquad (2.91)$$

with

$$H_0 = H_{part} + H_{rad}. \qquad (2.92)$$

As was demonstrated in Sect. 2.2 of this chapter, when there is no radiation field, the total system is made up of charged particles only, while when these sources vanish, there is only the free field. Hence H_0 represents a solved problem, and which is separable when the sub-systems do not interact. Known solutions to the atomic and molecular Hamiltonian are represented by $H_{part}(\xi)|E_m^\xi\rangle = E_m^\xi|E_m^\xi\rangle$, where $|E_m^\xi\rangle$ is the energy eigenket associated with energy eigenvalue E_m^ξ for species

ξ in electronic state represented by quantum number m, with additional labels being inserted to describe extra degrees of freedom. An occupation number state is used to specify the radiation field according to $H_{rad}|n(\vec{k}, \lambda)\rangle = (n + \frac{1}{2})\hbar ck|n(\vec{k}, \lambda)\rangle$, where the energy of the electromagnetic field is $E_{rad} = (n + \frac{1}{2})\hbar\omega$, when there are n photons of mode $(\vec{k}, \lambda)$. Other radiation field states, such as a coherent state representation [9], may be used in place of a number state specification. Hence the basis functions employed in the perturbation theory solution to Eq. (2.91) are product particle-radiation field states

$$|part\rangle|rad\rangle = |part; rad\rangle = \left|E_m^{\xi}; n(\vec{k}, \lambda)\right\rangle. \tag{2.93}$$

The total energy of state (2.93) is given by $E_m^{\xi} + n\hbar ck$.

The question often asked in any quantum mechanical problem, is given that the system at some initial time t_i is in state $|i\rangle$, what is the probability that it is in state $|f\rangle$ at some later time t_f, due to the influence of the perturbation, which may or may not be time-dependent, but acts during the time interval? With exact analytical solutions only possible for a few limited choices of H_{int}, a perturbation theory solution is developed for the probability amplitude M_{fi} in series of powers of H_{int} for the transition $|f\rangle \leftarrow |i\rangle$. The observable quantity commonly derived from the matrix element is the transition rate, Γ, associated with Fermi's golden rule [9],

$$\Gamma = \frac{2\pi}{\hbar}|M_{fi}|^2 \rho_f, \tag{2.94}$$

where ρ_f is the density of final states. Equation (2.94) holds in the weak-coupling regime. When the initial and final states are identical, corresponding to a diagonal matrix element, the resulting observable may be interpreted as an energy shift, which is especially useful in the chapters to follow when dispersion potentials are evaluated. In powers of the perturbation operator, we find for the perturbed energy

$$E_m = E_m^{(0)} + \left\langle m^{(0)}|H_{int}|m^{(0)}\right\rangle + \sum_{\substack{n \\ m \neq n}} \frac{\left\langle m^{(0)}|H_{int}|n^{(0)}\right\rangle\left\langle n^{(0)}|H_{int}|m^{(0)}\right\rangle}{E_m^{(0)} - E_n^{(0)}} + \cdots, \tag{2.95}$$

in terms of the unperturbed states and energies $|m^{(0)}\rangle$ and $E_m^{(0)}$, respectively, a sum of zeroth-, first-, second-, and higher-order terms in H_{int}.

References

1. Schwinger JS (ed) (1958) Selected papers on quantum electrodynamics. Dover, New York
2. Goldstein H (1960) Classical mechanics. Addison-Wesley, Reading, MA
3. Dirac PAM (1958) The principles of quantum mechanics. Clarendon, Oxford

4. Morse PM, Feshbach H (1953) Methods of theoretical physics. McGraw-Hill, New York
5. Jackson JD (1975) Classical electrodynamics. Wiley, New York
6. Power EA (1964) Introductory quantum electrodynamics. Longmans, London
7. Born M, Heisenberg W, Jordan P (1926) Zur quantenmechanik II. Z Phys 35:557
8. Jeans JH (1905) On the partition of energy between matter and aether. Philos Mag 10:91
9. Craig DP, Thirunamachandran T (1998) Molecular quantum electrodynamics. Dover, New York
10. Salam A (2010) Molecular quantum electrodynamics. Wiley, Hoboken
11. Born M, Oppenheimer JR (1927) Zur quantentheorie der molekeln. Ann Phys 84:457
12. Schrödinger E (1926) Der stetige Übergang von der Mikro-zur Makromechanik. Naturwissenschaften 14:664
13. Milonni PW (1994) The quantum vacuum. Academic Press, San Diego
14. Power EA (1978) A review of canonical transformations as they affect multiphoton processes. In: Eberly JH, Lambropoulos P (eds) Multiphoton processes. Wiley, New York, pp 11–46
15. Power EA, Zienau S (1959) Coulomb gauge in non-relativistic quantum electrodynamics and the shape of spectral lines. Phil Trans Roy Soc London A251:427
16. Woolley RG (1971) Molecular quantum electrodynamics. Proc Roy Soc London A321:557
17. Babiker M, Power EA, Thirunamachandran T (1974) On a generalisation of the Power-Zienau-Woolley transformation in quantum electrodynamics and atomic field equations. Proc Roy Soc London A338:235
18. Power EA, Thirunamachandran T (1980) The multipolar Hamiltonian in radiation theory. Proc Roy Soc London A372:265
19. Cohen-Tannoudji C, Dupont-Roc J, Grynberg G (1989) Photons and atoms. Wiley, New York
20. Healy WP (1982) Non-relativistic quantum electrodynamics. Academic Press, London
21. Compagno G, Passante R, Persico F (1995) Atom-field interactions and dressed atoms. Cambridge University Press, Cambridge
22. Göppert-Mayer M (1931) Über elementarakte mit zwei quantensprüngen. Ann Phys Leipzig 9:273
23. Power EA, Thirunamachandran T (1978) On the nature of the Hamiltonian for the interaction of radiation with atoms and molecules: (e/mc)p.A, $-\mu$.E, and all that. Am J Phys 46:370

Chapter 3
Dispersion Interaction Between Two Atoms or Molecules

Abstract In this chapter, diagrammatic time-dependent perturbation theory is employed to calculate the Casimir-Polder dispersion potential between two neutral electric dipole polarisable atoms or molecules. Its computation via the minimal-coupling scheme is summarised first. Next, it is shown how the energy shift may be computed more simply by adopting the multipolar Hamiltonian in the electric dipole approximation. In this second framework the force is mediated by the exchange of two virtual photons, and the Casimir-Polder formula results on summing the contribution from twenty-four time-ordered diagrams evaluated at fourth-order of perturbation theory. The potential is obtained for oriented as well as for isotropic systems separated beyond the region of wave function overlap. Asymptotically limiting forms of the interaction energy applicable in the near-and far-zone regions are also found. The former reproduces the London dispersion formula, while the latter exhibits an inverse seventh power law due to the effects of retardation.

Keywords Diagrammatic perturbation theory · Casimir-Polder potential · London dispersion energy · Vacuum electromagnetic field · Fluctuating electric dipoles

3.1 Casimir-Polder Potential: Minimal-Coupling Calculation

Even though Göppert-Mayer [1], and others, employed the electric dipole coupling Hamiltonian to study the interaction of light with matter in classical, semi-classical or fully quantal descriptions, in their seminal quantum electrodynamical calculation of the van der Waals dispersion potential, Casimir and Polder [2] used the conventional minimal-coupling framework. Here we briefly outline the main features of their computation and present the result for the energy shift, before showing in the following section how the potential may be obtained more straightforwardly using the multipolar-coupling scheme.

© The Author(s) 2016
A. Salam, *Non-Relativistic QED Theory of the van der Waals Dispersion Interaction*, SpringerBriefs in Electrical and Magnetic Properties of Atoms, Molecules, and Clusters, DOI 10.1007/978-3-319-45606-5_3

Consider two atoms or neutral, non-polar molecules A and B, both of which are in their ground electronic state. Let the two particles be positioned at $\vec{R}_A$ and $\vec{R}_B$, respectively, with relative separation distance $R = |\vec{R}_B - \vec{R}_A|$. The total Hamiltonian operator in the minimal-coupling formalism for the two particles and the electromagnetic field, and the interaction between them, is given by

$$H^{min} = \sum_{\xi=A,B} H_{part}(\xi) + H_{rad} + \sum_{\xi=A,B} H_{int}(\xi) + H_{int}(A,B), \tag{3.1}$$

where $H_{part}(\xi)$ is given by Eq. (2.43), and H_{rad} expressed in terms of electric and magnetic fields is

$$H_{rad} = \frac{\varepsilon_0}{2} \int \left\{ \vec{e}^{\perp 2}(\vec{r}) + c^2 \vec{b}^2(\vec{r}) \right\} d^3 \vec{r}. \tag{3.2}$$

The third term of Eq. (3.1) is

$$H_{int}(\xi) = -\sum_{\alpha} \frac{e_\alpha}{m_\alpha} \vec{p}_\alpha(\xi) \cdot \vec{a}(\vec{q}_\alpha(\xi) - \vec{R}_\xi) + \sum_{\alpha} \frac{e_\alpha^2}{2m_\alpha} \vec{a}^2(\vec{q}_\alpha(\xi) - \vec{R}_\xi). \tag{3.3}$$

In addition to fluctuations due to the electromagnetic vacuum, the material particles are sources of radiation. If the wavelength of the radiation engaging with each entity is much larger than atomic and molecular dimensions, corresponding to the long wavelength approximation, then spatial variations of the vector potential may be ignored. Retaining the first term in the Taylor series expansion of the vector potential results in the electric dipole approximation. If the dipoles are located at $\vec{R}_A$ and $\vec{R}_B$, then the two centre interaction term is the familiar static dipolar coupling

$$H_{int}(A,B) = \frac{\mu_i(A)\mu_j(B)}{4\pi\varepsilon_0 R^3} (\delta_{ij} - 3\hat{R}_i\hat{R}_j), \tag{3.4}$$

with the electric dipole moment operator defined by Eq. (2.88). Noting that the atomic polarisability is proportional to e^2, where $-e$ is the electronic charge, all contributions correct to order e^4 in the perturbation operator Eqs. (3.3) and (3.4) must therefore be accounted for in the calculation. Hence these will manifest in a variety of contributing orders in perturbation theory. There is a term in second order arising from $H_{int}(A,B)$. Diagrammatically it may be represented as shown in Fig. 3.1, in which the solid vertical lines depict the state of the atom or molecule, with time flowing upward, and the dashed horizontal lines indicate the instantaneous dipole-dipole interaction, Eq. (3.4), with this axis corresponding to the displacement variable. At second-order, two such couplings appear. As befits the dispersion force, both A and B are in the ground electronic state, with the electromagnetic field in the vacuum state. Transitions to virtual levels are possible, and these states in A and B are denoted by $|r\rangle$ and $|s\rangle$, respectively. As is well known,

Fig. 3.1 Contribution of
static dipolar coupling to
dispersion potential at
second-order of perturbation
theory

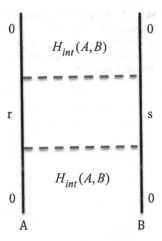

evaluation of the interaction using static dipolar coupling results in the London dispersion formula, exhibiting R^{-6} separation distance dependence [3], whose detailed form will be given in Sect. 3.3.

Another contribution at second-order arises from the second term of Eq. (3.3), that which is proportional to the square of the vector potential. It is illustrated in Fig. 3.2, in which two-photons are first simultaneously emitted at one site, and subsequently absorbed at the other centre. The two-photon interaction vertex is a consequence of the quadratic nature of the coupling, dependent upon the square of the vector potential. The two photons, which are unobservable and therefore termed *virtual* photons, are labelled by modes $(\vec{p}, \varepsilon)$ and $(\vec{p}', \varepsilon')$. Both real and virtual photons are sanctioned by theory, with virtual ones being invoked to interpret the mediation of coupling forces between particles, as occurs in quantum field theory. Because the polarisation and momenta of all virtual photons are summed over, the mode specifications are arbitrary, and serve merely as labels to distinguish between

Fig. 3.2 $\frac{e^2}{2m}\vec{a}^2(\vec{R}_\xi)$
contribution to the dispersion
potential occurring in
second-order of perturbation
theory

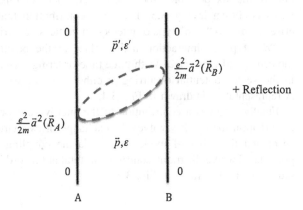

Fig. 3.3 One of six diagrams
involving "$\vec{p} \cdot \vec{a}$" and
$H_{int}(A, B)$ couplings
contributing to the dispersion
force at third-order

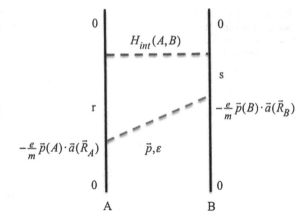

them in computational work, and in their visualisation by time-ordered graphical means. It is seen from Fig. 3.2 that for the specific interaction term being considered, that the contribution from two diagrams must be accounted for. This is because the pictorial representation of particle-electromagnetic field coupling events as shown in this figure is a variant of the space-time diagrams introduced by Feynman [4–6] to display and organise elementary photon absorption and emission events undertaken by electrons. The rules developed by him for evaluating contributions from diagrams in covariant QED also apply to photons interacting with non-relativistic charged particles [7]. To ensure that the correct result is arrived at for a specific process, be it the probability amplitude or the energy shift, all topologically distinct diagrams that connect the same initial and final states at a particular order of perturbation theory must be drawn and summed over.

There are two types of contributions occurring in third-order of perturbation theory. Exchange of a virtual photon of mode $(\vec{p}, \varepsilon)$ via the "$\vec{p} \cdot \vec{a}$" interaction term of Eq. (3.3), the single-photon emission and absorption events contributing two-orders overall, and the static coupling $H_{int}(A, B)$, which contributes one order. One of the six possible time-ordered sequences of this combination of coupling terms is shown in Fig. 3.3. The other contribution that is of third-order overall arises from two "$\vec{p} \cdot \vec{a}$" type of interaction vertices occurring at one particle site, and an "$\vec{a}^2$" type of interaction taking place at the position of the other atom or molecule, and *vice versa*. With three time-orderings possible for the term quadratic in the vector potential occurring at either A or B, six such diagrams contribute. A representative is drawn in Fig. 3.4.

Finally, there is a contribution to the energy shift occurring at fourth-order of perturbation theory. It involves the exchange of two virtual photons, one of mode $(\vec{p}, \varepsilon)$ and the other of mode $(\vec{p}', \varepsilon')$ via the coupling term linear in the vector potential. Twelve diagrams contribute in total at this order. One of the time-ordered sequences is illustrated in Fig. 3.5.

Fig. 3.4 One of six time-orderings involving the "$\vec{p} \cdot \vec{a}$" interaction at one centre and "$\vec{a}^2$" at the other in third-order of perturbation theory

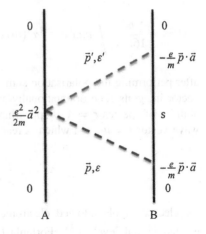

Fig. 3.5 One of twelve time-orderings involving "$\vec{p} \cdot \vec{a}$" type of couplings occurring in fourth-order of e

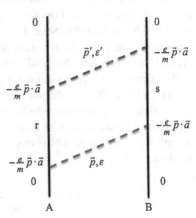

The contributions from all of the possible time-ordered sequences, representative diagrams of which were shown in Figs. 3.1, 3.2, 3.3, 3.4, 3.5, of which a greater number are drawn explicitly in Power's textbook [8], are evaluated individually and then summed, in a technically involved calculation. In order to arrive at an energy shift proportional to the atomic or molecular electric dipole polarisability of each species, position matrix elements are related to their conjugate momentum counterparts through the commutator connecting these variables, so that

$$E_{rs}\vec{q}^{\,rs} = \left\langle E_r \right| [H_{part}, \vec{q}] \left| E_s \right\rangle = \frac{\hbar}{im}\vec{p}^{\,rs}, \qquad (3.5)$$

where E_{rs} represents the difference in energy between states $|r\rangle$ and $|s\rangle$, $E_r - E_s$, and the rs-th matrix element of particle dynamical variable $\vec{X}$ is $\vec{X}^{rs} = \langle r|\vec{X}|s\rangle$.

For isotropic species, the van der Waals dispersion interaction energy evaluated by Casimir and Polder via the minimal-coupling QED Hamiltonian is [2]

$$\Delta E = -\frac{\hbar c}{16\pi^3 \varepsilon_0^2} \int\limits_0^\infty duu^6 e^{-2uR}\alpha^A(iu)\alpha^B(iu)\left[\frac{1}{u^2 R^2}+\frac{2}{u^3 R^3}+\frac{5}{u^4 R^4}+\frac{6}{u^5 R^5}+\frac{3}{u^6 R^6}\right], \quad (3.6)$$

after performing the polarisation sums (see below) and evaluating one of the wave vector integrals. The orientationally averaged ground state electric dipole polarisabilities of species $\xi = A, B$ appearing in Eq. (3.6) are evaluated at the imaginary wave vector $k = iu$, and which is readily obtained from its usual expression

$$\alpha^\xi(k) = \frac{2}{3}\sum_t \frac{\left|\vec{\mu}^{0t}(\xi)\right|^2 E_{t0}^\xi}{\left(E_{t0}^\xi\right)^2 - (\hbar ck)^2}, \quad (3.7)$$

for electric dipole allowed electronic transitions from the ground state $\left|0^\xi\right\rangle$ to excited virtual levels $\left|t^\xi\right\rangle$. Formula (3.6) holds for all pair separation distances R beyond the contact and overlap region of A and B. Limiting forms of the general result will be evaluated in Sect. 3.3 after details of the computation in the multipolar framework are given in the next section. As expected, the dispersion potential depends on the polarisability of each interacting particle.

The functional form (3.6) properly includes the effects of retardation as will be shown below. This is despite the fact that the calculation was done in the minimal-coupling scheme, which explicitly contains a static inter-particle interaction term. Further insight in relation to this aspect is gained by noting that even though the vector potential obeys the wave equation in the Coulomb gauge with solution

$$\vec{a}(\vec{r},t) = \frac{1}{4\pi\varepsilon_0 c^2}\int \frac{\vec{j}^\perp(\vec{r}',t-|\vec{r}-\vec{r}'|/c)}{|\vec{r}-\vec{r}'|}d^3\vec{r}', \quad (3.8)$$

$\vec{a}^\perp(\vec{r},t)$ is a non-local vector and is not fully retarded [9]. This is because its source is not the total current but its transverse component,

$$j_i^\perp(\vec{r},t) = \int j_j(\vec{r}',t)\delta_{ij}^\perp(\vec{r}-\vec{r}')d^3\vec{r}', \quad (3.9)$$

which is also a non-local vector. With $\vec{\Pi}(\vec{r}) = -\varepsilon_0\vec{e}^\perp(\vec{r}) = \varepsilon_0\dot{\vec{a}}(\vec{r})$ in minimal-coupling, we see that $\vec{e}^\perp(\vec{r},t)$ contains static terms. For instance, the transverse electric field of an electric dipole source oscillating at frequency $\omega = ck$, is [10]

$$e_i^\perp(\vec{\mu};\vec{r},t) = \frac{1}{4\pi\varepsilon_0}\mu_j e^{-i\omega t}(-\nabla^2\delta_{ij}+\nabla_i\nabla_j)\frac{e^{ikr}-1}{r}, \quad r<ct. \quad (3.10)$$

Note that the longitudinal electric field associated with this multipole moment is

$$e_i^{\parallel}(\vec{\mu};\vec{r},t) = \varepsilon_0^{-1} p_i^{\perp}(\vec{\mu};\vec{r},t) = \varepsilon_0^{-1}\mu_j e^{-i\omega t}\delta_{ij}^{\perp}(\vec{r}) = -\frac{1}{4\pi\varepsilon_0 r^3}\mu_j e^{-i\omega t}(\delta_{ij} - 3\hat{r}_i\hat{r}_j), \; r > 0,$$

$$(3.11)$$

on employing the form of the transverse delta function dyadic outside of the source. A retarded result is obtained after adding the longitudinal electric field, $\vec{e}^{\parallel}$ to the transverse component, $\vec{e}^{\perp}$ to give the total electric field,

$$e_i^{tot}(\vec{\mu};\vec{r},t) = e_i^{\perp}(\vec{\mu};\vec{r},t) + e_i^{\parallel}(\vec{\mu};\vec{r},t) = \frac{1}{4\pi\varepsilon_0}\mu_j(-\nabla^2\delta_{ij} + \nabla_i\nabla_j)\frac{e^{ik(r-ct)}}{r}$$
$$= \varepsilon_0^{-1} d_i^{\perp}(\vec{\mu};\vec{r},t). \qquad (3.12)$$

From the last equality of Eq. (3.12) it is seen that the total electric field is equal to ε_0^{-1} times the transverse component of the electric displacement field. It may be recalled that the latter is, to within a sign, the momentum field canonically conjugate to the vector potential in the multipolar formalism [9]. Interestingly, neither $\vec{e}^{\perp}$ nor $\vec{e}^{tot}$ vanish for $t < r/c$ in the minimal-coupling scheme, and have genuine acausal signals associated with them [11, 12]. Remarkably, there is exact cancellation of all static terms in the evaluation of the Casimir-Polder potential using H^{min}, leading to a causal result.

3.2   Casimir-Polder Energy Shift: Multipolar Formalism Calculation

Employing the multipolar-coupling QED Hamiltonian instead of the minimal-coupling version simplifies the computation of the dispersion potential in a significant way. From H^{mult} given by Eq. (2.75), after dropping the one-centre transverse electric polarisation field squared contribution, the total system Hamiltonian for two particles in the presence of an electromagnetic field is written as

$$H = H_{part}^{mult}(A) + H_{part}^{mult}(B) + H_{rad} + H_{int}^{mult}(A) + H_{int}^{mult}(B). \qquad (3.13)$$

Invoking the electric dipole approximation by retaining the first term of the perturbation operator Eq. (2.87), the last two terms of Eq. (3.13) are given by

$$\sum_{\xi=A,B} H_{int}^{mult}(\xi) = -\varepsilon_0^{-1}\vec{\mu}(A) \cdot \vec{d}^{\perp}(\vec{R}_A) - \varepsilon_0^{-1}\vec{\mu}(B) \cdot \vec{d}^{\perp}(\vec{R}_B). \qquad (3.14)$$

Closer inspection of the Feynman diagrams drawn in the previous section to aid the calculation of the dispersion energy shift in the minimal-coupling scheme

Fig. 3.6 One of twelve
diagrams contributing to the
Casimir-Polder potential
when evaluated in the
multipolar coupling
framework

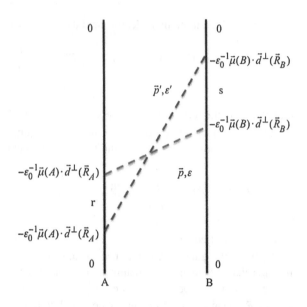

reveals the two particles A and B to be coupled by the exchange of two virtual photons through either "$\vec{p} \cdot \vec{a}$" or "$\vec{a}^2$", or a combination of these two interaction terms, or by exchange of one virtual photon and one static dipolar coupling interaction via "$\vec{p} \cdot \vec{a}$" and an $H_{int}(A, B)$ coupling, or through the action of $H_{int}(A, B)$ twice. With the form of interaction Hamiltonian in the multipolar framework given by Eq. (3.14), the dispersion force may be viewed in this formalism as arising from the exchange of two virtual photons between the pair. At each centre the coupling is therefore proportional to e^2. To leading order in R, the interaction energy for coupling between nonpolar molecules is calculated using the fourth-order perturbation theory formula

$$\Delta E = -\sum_{I,II,III} \frac{\langle 0|H_{int}|III\rangle\langle III|H_{int}|II\rangle\langle II|H_{int}|I\rangle\langle I|H_{int}|0\rangle}{(E_{III} - E_0)(E_{II} - E_0)(E_I - E_0)}, \qquad (3.15)$$

with interaction Hamiltonian given by Eq. (3.14). Since the coupling Hamiltonian is linear in the electric displacement field, creating or destroying a single photon, there are twenty-four possible time-ordered sequences of emission and absorption of two virtual photons, one of mode $(\vec{p}, \varepsilon)$, the other $(\vec{p}', \varepsilon')$. With virtual photons ultimately being indistinguishable, the number of diagrams is halved to avoid double counting. One of the twelve is sketched in Fig. 3.6.

The initial and final total system states are represented by the ket

$$|0\rangle = \left|E_0^A, E_0^B; 0(\vec{p}, \varepsilon), 0(\vec{p}', \varepsilon')\right\rangle, \qquad (3.16)$$

corresponding to both species in their electronic ground states, and the radiation field in the vacuum state, in which there are no photons present. The state $|0\rangle$ is easily read-off from the graph shown in Fig. 3.6 and other contributing diagrams, corresponding to the state of the system at initial and final times. Similarly, the intermediate particle-field states $|I\rangle, |II\rangle$, and $|III\rangle$ appearing in Eq. (3.15) are also readily written down from each graph after the respective photonic event has taken place. Sums are executed over all such intermediate states in order to arrive at an expression for the energy shift as required by formula (3.15). The denominator of Eq. (3.15) contains differences in energy between intermediate states and the ground state (3.16), which are also obtained straightforwardly from the respective diagram, or from explicit representation of the relevant intermediate state. As before, $|r\rangle$ and $|s\rangle$ denote virtual electronic states of A and B, respectively.

Evaluating the numerator and denominator of Eq. (3.15) for each graph and summing over all twelve diagrams yields for the interaction energy the intermediate expression [9, 13, 14]

$$\Delta E = -\sum_{\vec{p},\vec{p}'}\sum_{\varepsilon,\varepsilon'}\sum_{r,s}\frac{1}{\hbar c}\left(\frac{1}{\varepsilon_0 V}\right)^2 e_i^{(\varepsilon)}(\vec{p})\bar{e}_k^{(\varepsilon)}(\vec{p})e_j^{(\varepsilon')}(\vec{p}')\bar{e}_l^{(\varepsilon')}(\vec{p}')pp'e^{i(\vec{p}+\vec{p}')\cdot\vec{R}}$$

$$\times \mu_i^{0r}(A)\mu_j^{r0}(A)\mu_k^{0s}(B)\mu_l^{s0}(B)\frac{(k_{r0}+k_{s0}+p)}{(k_{r0}+k_{s0})(k_{r0}+p)(k_{s0}+p)}\left(\frac{1}{p+p'}-\frac{1}{p-p'}\right).$$

$$(3.17)$$

To proceed further, summation over virtual photon modes has to be performed. Noting that the two directions of polarisation and the wave vector may be represented by a set of three mutually perpendicular complex unit vectors $\vec{e}^{(1)}(\vec{p}), \vec{e}^{(2)}(\vec{p})$ and $\hat{p}$, the following identity is easily established,

$$\sum_{\varepsilon=1,2} e_i^{(\varepsilon)}(\vec{p})\bar{e}_j^{(\varepsilon)}(\vec{p}) = \delta_{ij} - \hat{p}_i\hat{p}_j, \qquad (3.18)$$

enabling the polarisation sums in Eq. (3.17) to be carried out immediately. If the dimensions of the box confining the electromagnetic field are large enough to quantify the number of lattice points in reciprocal space, then the wave vector sum over discrete $\vec{p}$-values may be converted to an integral via the substitution

$$\lim_{V\to\infty}\frac{1}{V}\sum_{\vec{p}} \to \left(\frac{1}{2\pi}\right)^3 \int d^3\vec{p}. \qquad (3.19)$$

The infinitesimal volume element $d^3\vec{p}$ may be expressed in a suitable system of coordinates, for example in Cartesians as $dp_x dp_y dp_z$. More convenient is its representation in spherical polar coordinates as $p^2 dp d\Omega$, where $d\Omega = \sin\theta d\theta d\phi$, with θ the polar angle and ϕ the azimuthal angle. Factors of the form

$$\int (\delta_{ij} - \hat{p}_i\hat{p}_j)e^{\pm i\vec{p}\cdot\vec{R}}d\Omega, \tag{3.20}$$

then occur, and the angular integration is performed on making use of the elementary integral

$$\frac{1}{4\pi}\int e^{\pm i\vec{p}\cdot\vec{R}}d\Omega = \frac{\sin pR}{pR}. \tag{3.21}$$

Effecting action of the gradient operator twice on Eq. (3.21) establishes the useful relation

$$\frac{1}{4\pi}\int (\delta_{ij} - \hat{p}_i\hat{p}_j)e^{\pm i\vec{p}\cdot\vec{R}}d\Omega = \operatorname{Im} F_{ij}(pR), \tag{3.22}$$

where the tensor field $F_{ij}(pR)$ is defined as

$$
\begin{aligned}
F_{ij}(pR) &= (-\nabla^2\delta_{ij} + \nabla_i\nabla_j)\frac{e^{ipR}}{p^3R} \\
&= \left\{ (\delta_{ij} - \hat{R}_i\hat{R}_j)\frac{1}{pR} + (\delta_{ij} - 3\hat{R}_i\hat{R}_j)\left(\frac{i}{p^2R^2} - \frac{1}{p^3R^3}\right) \right\}e^{ipR}.
\end{aligned} \tag{3.23}
$$

Performing the polarisation sums using Eq. (3.18), followed by application of Eq. (3.19) and the angular average Eq. (3.22), and evaluating the p'-integral in Eq. (3.17) at the pole $p' = -p$ treated as the principal value after extending the limits of integration from $[0, \infty]$ to $[-\infty, \infty]$ on noting that $\operatorname{Im}\left[F_{jl}(p'R)\right]$ is an even function of p', the energy shift becomes

$$
\begin{aligned}
\Delta E = &-\frac{1}{4\pi^3\hbar c\varepsilon_0^2}\sum_{r,s}\mu_i^{0r}(A)\mu_j^{r0}(A)\mu_k^{0s}(B)\mu_l^{s0}(B)\frac{1}{(k_{r0}+k_{s0})} \\
&\times \int_0^\infty dp\,p^6\,\frac{(k_{r0}+k_{s0}+p)}{(k_{r0}+p)(k_{s0}+p)}\operatorname{Re}\left[F_{jl}(pR)\right]\operatorname{Im}\left[F_{ik}(pR)\right],
\end{aligned} \tag{3.24}
$$

on using the result

$$\int_{-\infty}^\infty \frac{p'^3}{p+p'}\operatorname{Im}\left[F_{jl}(p'R)\right]dp' = \pi p^3\operatorname{Re}\left[F_{jl}(pR)\right]. \tag{3.25}$$

Calculating the product $\operatorname{Re}\left[F_{jl}(pR)\right]\operatorname{Im}\left[F_{ik}(pR)\right]$ using Eq. (3.23) and transforming the integral to the complex plane on introducing $p = \pm iu$, finally yields the Casimir-Polder potential for two species in fixed relative orientation,

$$\Delta E = -\frac{1}{8\pi^3 \varepsilon_0^2 \hbar c} \sum_{r,s} \mu_i^{0r}(A) \mu_j^{r0}(A) \mu_k^{0s}(B) \mu_l^{s0}(B) \int\limits_0^\infty duu^6 e^{-2uR} \frac{k_{r0}k_{s0}}{(k_{r0}^2 + u^2)(k_{s0}^2 + u^2)}$$

$$\times \left[\begin{array}{l} \frac{(\delta_{ik}-\hat{R}_i\hat{R}_k)(\delta_{jl}-\hat{R}_j\hat{R}_l)}{u^2 R^2} + \frac{(\delta_{ik}-\hat{R}_i\hat{R}_k)(\delta_{jl}-3\hat{R}_j\hat{R}_l)+(\delta_{ik}-3\hat{R}_i\hat{R}_k)(\delta_{jl}-\hat{R}_j\hat{R}_l)}{u^3 R^3} \\ + \frac{(\delta_{ik}-\hat{R}_i\hat{R}_k)(\delta_{jl}-3\hat{R}_j\hat{R}_l)+(\delta_{ik}-3\hat{R}_i\hat{R}_k)(\delta_{jl}-\hat{R}_j\hat{R}_l)+(\delta_{ik}-3\hat{R}_i\hat{R}_k)(\delta_{jl}-3\hat{R}_j\hat{R}_l)}{u^4 R^4} \\ + \frac{2(\delta_{ik}-3\hat{R}_i\hat{R}_k)(\delta_{jl}-3\hat{R}_j\hat{R}_l)}{u^5 R^5} + \frac{(\delta_{ik}-3\hat{R}_i\hat{R}_k)(\delta_{jl}-3\hat{R}_j\hat{R}_l)}{u^6 R^6} \end{array} \right].$$

$$(3.26)$$

To obtain the potential for isotropic particles, a rotational average is carried out using the result [15]

$$\left\langle \mu_i^{0t}(\xi) \mu_j^{t0}(\xi) \right\rangle = \frac{1}{3} \delta_{ij} \delta_{\lambda\mu} \mu_\lambda^{0t}(\xi) \mu_\mu^{t0}(\xi) = \frac{1}{3} \delta_{ij} \left| \vec{\mu}^{0t}(\xi) \right|^2, \qquad (3.27)$$

where Greek subscripts designate Cartesian tensor components in the molecule-fixed frame of reference. Contracting the tensor products with the factors within square brackets of Eq. (3.26), and using the polarisability at imaginary frequency from Eq. (3.7), yields expression (3.6).

3.3   Asymptotically Limiting Forms

From the Casimir-Polder potential, Eq. (3.6), which holds for all pair separation distances outside regions of wave function overlap, the forms of the interaction energy at short and large R are readily obtained. In the former limit, the inter-particle displacement is small relative to characteristic reduced transition wavelengths in either species, that is $k_{r0}R$ and $k_{s0}R$ are both considerably less than unity. Hence the dominant term within square brackets of Eq. (3.6) is the R^{-6} dependent one. Approximating the exponential factor by unity, and using the standard integral

$$\int\limits_0^\infty \frac{rsdu}{(r^2 + u^2)(s^2 + u^2)} = \frac{\pi}{2} \frac{1}{(r+s)}, \quad r,s > 0, \qquad (3.28)$$

results in the near-zone (NZ) asymptotically limiting interaction energy,

$$\Delta E^{NZ} = -\frac{1}{24\pi^2 \varepsilon_0^2 R^6} \sum_{r,s} \frac{\left| \vec{\mu}^{0r}(A) \right|^2 \left| \vec{\mu}^{s0}(B) \right|^2}{E_{r0}^A + E_{s0}^B}. \qquad (3.29)$$

Expression (3.29) is the dispersion energy formula of London [3]. It is interesting that when the particles are very close to one another, coupling between them

Fig. 3.7 Four diagrams which yield the London dispersion potential

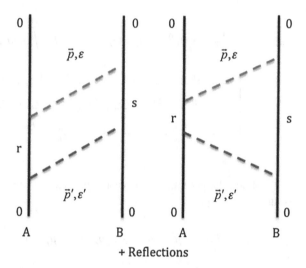

+ Reflections

may be taken to be effectively instantaneous in action. As already pointed out, Eq. (3.29) results on evaluating the static dipolar coupling potential between two ground state species at second-order of perturbation theory.

Another way to directly arrive at Eq. (3.29) is to note that when R is small, the virtual photons traversing the pair are propagating for a short time only, and are therefore highly energetic, as demanded by Heisenberg's uncertainty principle. In the present example virtual photon wave vectors p and p' will be very much greater than k_{r0} and k_{s0}. Terms whose denominators may be approximated by $pp'(k_{r0} + k_{s0})$ will therefore make the largest contribution to the interaction energy. These arise from diagrams in which one virtual photon is first emitted and absorbed, followed by the second one. Four of the twelve possible time-orderings meet this requirement. They are illustrated in Fig. 3.7. Summation of the contributions from these four graphs gives Eq. (3.29). It now becomes clear how results embodying static coupling may be extracted from retarded potentials and how the diagram of Fig. 3.1 follows from those drawn in Fig. 3.7 to yield the London formula.

Physical arguments opposite to those presented for the near-zone may be applied to obtain the asymptotically limiting form of the Casimir-Polder dispersion potential at very long-range, where retardation effects dominate. At this extreme, the pair separation distance is much greater than characteristic reduced transition wavelengths, namely $k_{r0}R \gg 1$ and $k_{s0}R \gg 1$. This means that u^2 may be discarded in the energy denominators of the atomic or molecular polarisabilities appearing in Eq. (3.6), and the five subsequent u-integrals may be evaluated individually using

$$\int_0^\infty x^n e^{-\eta x}dx = n!\eta^{-n-1}, \quad \mathrm{Re}\,\eta > 0, \qquad (3.30)$$

and summed, to give the far-zone (FZ) limit

$$\Delta E^{FZ} = -\frac{23\hbar c}{64\pi^3 \varepsilon_0^2 R^7} \alpha^A(0)\alpha^B(0), \tag{3.31}$$

exhibiting inverse seventh power dependence on R. To be effective at large R, the exchanged virtual photons must be in existence for a comparatively long time, hence they are of low energy. This accounts for the polarisabilities appearing in Eq. (3.31) being static, corresponding to the zero frequency limit $\omega = ck = icu \to 0$. This atomic or molecular quantity is readily obtained from Eq. (3.7) and is given by

$$\alpha^\xi(0) = \frac{2}{3}\sum_t \frac{|\vec{\mu}^{0t}(\xi)|^2}{E_{t0}^\xi}. \tag{3.32}$$

With E_{r0} and E_{s0} both clearly being much greater than virtual photon energies $\hbar cp$ and $\hbar cp'$, of the twelve energy denominators contributing to the Casimir-Polder potential [9, 13], a sub-set of four approximated as $E_{r0}E_{s0}(\hbar cp + \hbar cp')$ will make the largest contribution in the far-zone region. They arise from diagrams in which the two virtual photons are simultaneously in transit when intermediate state $|II\rangle$ is evaluated in the fourth-order perturbation theory formula. There are four such orderings. One of these diagrams was shown in Fig. 3.6. The other is drawn in Fig. 3.8. The two remaining diagrams are reflections of Figs. 3.6 and 3.8. Summing the four graphs produces Eq. (3.31).

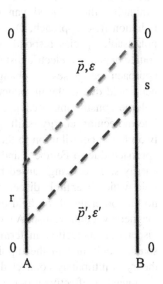

Fig. 3.8 One of four graphs contributing to the far-zone dispersion energy shift

Finally, it is worth remarking that Craig and Power [16] showed that the wave-
or far-zone limit may be obtained via second-order perturbation theory using an
effective interaction Hamiltonian that is directly proportional to the static electric
dipole polarisability and which has a quadratic dependence on the electric displacement field,

$$H_{int}^{eff} = -\frac{1}{2\varepsilon_0^2} \sum_{modes} \alpha_{ij}^A(0) d_i^\perp(\vec{R}_A) d_j^\perp(\vec{R}_A) - \frac{1}{2\varepsilon_0^2} \sum_{modes} \alpha_{kl}^B(0) d_k^\perp(\vec{R}_B) d_l^\perp(\vec{R}_B). \quad (3.33)$$

The four diagrams are now reduced to two diagrams of the type illustrated in
Fig. 3.2 but with coupling Hamiltonian (3.33) instead of the "$\vec{a}^2$" type of vertex.
The linear interaction vertices of the form $-\varepsilon_0^{-1}\vec{\mu}(\xi) \cdot \vec{d}^\perp(\vec{R}_\xi)$, which create or
destroy a single photon, have been collapsed, giving rise to two-photon coupling
vertices. The interaction Hamiltonian (3.33) is advantageous, especially when
computing higher multipole moment contributions to the dispersion force and
interactions between three-bodies, as will be shown in what follows, when the
number of time-ordered diagrams that have to be summed over can become pro-
hibitively large if an H_{int} that is linear in the Maxwell field or fields is employed.

3.4   Correlation of Fluctuating Electric Dipoles

There are a number of other physical viewpoints and computational schemes for
evaluating the retarded van der Waals dispersion potential within a quantised
radiation field approach. One of these is response theory in which an electrically
polarisable species responds via its dynamic polarisability to the electric dipole
source dependent electric displacement field of a second particle [17–19]. A second
approach recognises the change in the mode structure of the vacuum electromagnetic field due to the introduction of one or more atoms or molecules in the finite
volume constituting a cavity. The presence of two such objects gives rise to an
intermolecular energy shift [20–22]. A third involves the calculation of the
two-photon contribution to the elastic scattering amplitude in a model independent
approach due to Feinberg and Sucher [23]. A fourth is the re-interpretation of the
energy shift as being caused by the radiation reaction field instead of the vacuum
field, which is entirely dispensed with [24], and which may be accomplished with a
normal ordering of the field operators, whose expectation values over the electromagnetic vacuum vanish. A closely related fifth approach is the manifestation of the
dispersion interaction in terms of coupling of induced dipoles arising from the
source fields due to the dipoles themselves [25], which in the context of QED
theory is intimately connected to the source theory of Schwinger et al. [26], who
calculated an effective product of polarisation sources, from which the inter-particle
energy shift was obtained. The effect of a second dipole due to the electric field of a
first dipole was also used by Spruch and Kelsey [27] to compute the dispersion

potential. A sixth method is the calculation of the electromagnetic energy density produced by a polarisable source and the potential that ensues when a second body is in close proximity, giving rise to a mutual force of attraction [28].

As an alternative to the diagrammatic perturbation theory calculation detailed in earlier sections of this chapter, and those mentioned above, we adopt a picture in which quantum fluctuations of the vacuum electromagnetic field induce dipole moments in each of the two species. Their coupling then gives rise to an interaction energy, which is shown to be identical to the Casimir-Polder dispersion potential [29, 30]. Historically the method is of some importance as it was the viewpoint originally taken by London in his derivation of the R^{-6} functional form of the dispersion energy shift.

Consider a neutral polarisable species ξ located at $\vec{R}_\xi$. Its frequency dependent electric dipole polarisability is defined as

$$\alpha_{ij}^\xi(\omega) = \sum_t \left\{ \frac{\mu_i^{0t}(\xi)\mu_j^{t0}(\xi)}{E_{t0} - \hbar\omega} + \frac{\mu_j^{0t}(\xi)\mu_i^{t0}(\xi)}{E_{t0} + \hbar\omega} \right\}, \qquad (3.34)$$

in terms of the $t \leftarrow 0$ electric dipole allowed transition in ξ. Application of an electric displacement field of mode $(\vec{k}, \lambda)$ and frequency $\omega = ck$, induces an electric dipole moment

$$\mu_i^{ind}(\xi; \omega) = \varepsilon_0^{-1}\alpha_{ij}^\xi(\omega)d_j^\perp(\omega; \vec{R}_\xi). \qquad (3.35)$$

For two interacting atoms or molecules A and B, the moments induced at each centre couple according to

$$\Delta E = \sum_{modes} \mu_i^{ind}(A; \omega)\mu_j^{ind}(B; \omega)\text{Re } V_{ij}(\omega, \vec{R}), \qquad (3.36)$$

giving rise to an energy shift, with the separation distance defined as before, and the sum is taken over all radiation field modes. Appearing in Eq. (3.36) is the familiar retarded dipole-dipole coupling tensor [14],

$$V_{ij}(\omega, \vec{R}) = \frac{1}{4\pi\varepsilon_0 R^3} \left[(\delta_{ij} - 3\hat{R}_i\hat{R}_j)(1 - i\omega R/c) - (\delta_{ij} - \hat{R}_i\hat{R}_j)\omega^2 R^2/c^2 \right] e^{i\omega R/c}. \qquad (3.37)$$

Substituting the induced moments from Eq. (3.35) into (3.36) produces an expression that is proportional to the polarisability of each species,

$$\Delta E = \sum_{modes} \varepsilon_0^{-2}\alpha_{ik}^A(\omega)\alpha_{jl}^B(\omega)d_k^\perp(\omega; \vec{R}_A)d_l^\perp(\omega; \vec{R}_B)\text{Re } V_{ij}(\omega, \vec{R}), \qquad (3.38)$$

and consists of separate particle and field factors. To evaluate the van der Waals dispersion interaction energy, the expectation value of Eq. (3.38) is taken over the total system ground state Eq. (3.16). In this computational scheme, the induction of moments due to fluctuations of the vacuum electromagnetic field is readily apparent. Calculation of the expectation value of the molecular factors in Eq. (3.38) yields ground state dynamic polarisabilities of each species of the form Eq. (3.34), with $|r\rangle$ and $|s\rangle$ being virtual states of A and B as before. For the factor involving the product of electric displacement fields evaluated at two different positions, corresponding to the locations of the two particles, we find

$$\langle 0(\vec{p}, \varepsilon)| \sum_{modes} d_i^{\perp}(\omega; \vec{R}_A) d_j^{\perp}(\omega; \vec{R}_B) |0(\vec{p}, \varepsilon)\rangle = \frac{\hbar c \varepsilon_0}{16\pi^3} \int (\delta_{ij} - \hat{p}_i \hat{p}_j) p^3 e^{-i\vec{p} \cdot \vec{R}} dp d\Omega,$$

(3.39)

on using the mode expansion for the electric displacement field Eq. (2.73), carrying out the polarisation sum via Eq. (3.18), and on employing Eq. (3.19) to convert the sum over $\vec{p}$ to an integral. Performing the angular integration via Eq. (3.22), the interaction energy Eq. (3.38) after inserting Eq. (3.39) becomes

$$\Delta E = -\frac{\hbar c}{16\pi^3 \varepsilon_0^2 R^2} \int_0^{\infty} dp p^6 \alpha_{ik}^A(p) \alpha_{jl}^B(p) \operatorname{Re}[F_{ij}(pR)] \operatorname{Im}[F_{kl}(pR)],$$

(3.40)

when expressed in terms of the tensor Eq. (3.23), noting that $4\pi\varepsilon_0 V_{ij}(p, \vec{R}) = -p^3 F_{ij}(pR)$. Multiplying the last two factors in Eq. (3.40) and contracting the Cartesian tensor products after performing an orientational average using result Eq. (3.27) produces,

$$\Delta E = -\frac{\hbar c}{16\pi^3 \varepsilon_0^2 R^2} \int_0^{\infty} dp p^4 \alpha^A(p) \alpha^B(p) \left[\sin 2pR \left(1 - \frac{5}{p^2 R^2} + \frac{3}{p^4 R^4} \right) + \cos 2pR \left(\frac{2}{pR} - \frac{6}{p^3 R^3} \right) \right].$$

(3.41)

On substituting the imaginary variable $p = iu$ and transforming the integral to one over the complex frequency plane yields the Casimir-Polder potential Eq. (3.6), on using the analytic properties of the polarisability Eq. (3.34).

This method of coupling fluctuating electric dipole moments induced by the electric displacement field has been extended to treat the case in dispersion when one of the species is electronically excited [29], in which a downward transition occurs via emission of a real photon, and to the computation of the ground state dispersion potential between two chiral molecules [30].

In this chapter it has been shown how molecular QED may be employed to calculate the retarded van der Waals dispersion potential between a pair of neutral species in their ground electronic states. Diagrammatic time-dependent perturbation theory with either the minimal- or multipolar-coupling Hamiltonians yields the Casimir-Polder energy shift. The same result is found on viewing the interaction energy as arising from the coupling of electric dipole moments induced at each site by the vacuum electromagnetic field.

References

1. Göppert-Mayer M (1931) Über elementarakte mit zwei quantensprüngen. Ann Phys Leipzig 9:273
2. Casimir HBG, Polder D (1948) The influence of retardation on the London van der Waals forces. Phys Rev 73:360
3. London F (1937) The general theory of molecular forces. Trans Faraday Soc 33:8
4. Feynman RP (1949) The theory of positrons. Phys Rev 76:749
5. Feynman RP (1949) Space-time approach to quantum electrodynamics. Phys Rev 76:769
6. Schweber SS (1986) Feynman and the visualisation of space-time processes. Rev Mod Phys 58:449
7. Ward JF (1965) Calculation of nonlinear optical susceptibilities using diagrammatic perturbation theory. Rev Mod Phys 37:1
8. Power EA (1964) Introductory Quantum Electrodynamics. Longmans, London
9. Craig DP, Thirunamachandran T (1998) Molecular Quantum Electrodynamics. Dover, New York
10. Power EA, Thirunamachandran T (1983) Quantum electrodynamics with nonrelativistic sources. II. Maxwell fields in the vicinity of an atom. Phys Rev A 28:2663
11. Salam A (1997) Maxwell field operators, the energy density, and the Poynting vector calculated using the minimal-coupling framework of molecular quantum electrodynamics in the Heisenberg picture. Phys Rev A 56:2579
12. Power EA, Thirunamachandran T (1999) Time dependence of operators in minimal and multipolar nonrelativistic quantum electrodynamics. I. Electromagnetic fields in the neighbourhood of an atom. Phys Rev A 60:4927
13. Alligood BW, Salam A (2007) On the application of state sequence diagrams to the calculation of the Casimir-Polder potential. Mol Phys 105:395
14. Salam A (2010) Molecular Quantum Electrodynamics. John Wiley & Sons Inc, Hoboken
15. Andrews DL, Thirunamachandran T (1977) On three-dimensional rotational averages. J Chem Phys 67:5026
16. Craig DP, Power EA (1969) The asymptotic Casimir-Polder potential from second-order perturbation theory and its generalisation for anisotropic polarisabilities. Int J Quant Chem 3:903
17. Power EA, Thirunamachandran T (1983) Quantum electrodynamics with nonrelativistic sources. III. Intermolecular interactions. Phys Rev A 28:2671
18. Power EA, Thirunamachandran T (1993) Quantum electrodynamics with nonrelativistic sources. V. Electromagnetic field correlations and intermolecular interactions between molecules in either ground or excited states. Phys Rev A 47:2539
19. Salam A (2008) Molecular quantum electrodynamics in the Heisenberg picture: a field theoretic viewpoint. Int Rev Phys Chem 27:405
20. Casimir HBG (1949) Sur les forces van der Waals-London. J Chim Phys 46:407
21. Milonni PW (1994) The Quantum Vacuum. Academic Press, San Diego
22. Power EA, Thirunamachandran T (1994) Zero-point energy differences and many-body dispersion forces. Phys Rev A 50:3929
23. Feinberg G, Sucher J (1970) General theory of the van der Waals interaction: a model-independent approach. Phys Rev A 2:2395

24. Milonni PW (1982) Casimir forces without the vacuum radiation field. Phys Rev A 25:1315
25. Milonni PW, Shih M-L (1992) Source theory of Casimir force. Phys Rev A 45:4241
26. Schwinger JS, DeRaad LL Jr, Milton KA (1978) Casimir effect in dielectrics. Ann Phys 115:1 (NY)
27. Spruch L, Kelsey EJ (1978) Vacuum fluctuation and retardation effects on long-range potentials. Phys Rev A 18:845
28. Compagno G, Passante R, Persico F (1983) The role of the cloud of virtual photons in the shift of the ground-state energy of a hydrogen atom. Phys Lett A 98:253
29. Power EA, Thirunamachandran T (1993) Casimir-Polder potential as an interaction between induced dipoles. Phys Rev A 48:4761
30. Craig DP, Thirunamachandran T (1999) New approaches to chiral discrimination in coupling between molecules. Theo Chem Acc 102:112

Chapter 4
Inclusion of Higher Multipole Moments

Abstract In this chapter, contributions to the retarded two-body dispersion potential arising from higher multipole moments, are computed. First, a general formula is obtained for the energy shift between two species with pure electric multipole polarisability of arbitrary order. Interaction energies between an electric dipole polarisable molecule and a second that is electric quadrupole or octupole polarisable, as well as between two electric quadrupole polarisable molecules, are then extracted. Useful insight is gained by decomposing the octupole moment into irreducible components of weights-1 and -3. Magnetic effects are accounted for by including magnetic dipole and lowest-order diamagnetic coupling terms in the interaction Hamiltonian in addition to the electric dipole contribution. Remarkably, the energy shift between an electrically polarisable atom and a paramagnetically susceptible one is repulsive, unlike earlier examples, which are always attractive when both species are in the ground electronic state. Another interesting case arises when both molecules are chiral, and characterised by mixed electric-magnetic dipole polarisability. The energy shift is found to be discriminatory, dependent upon the handedness of each species. Asymptotic forms for ΔE are obtained for each of the examples considered. In all of the cases involving electric coupling, retardation effects weaken the potentials relative to their semi-classically derived counterparts.

Keywords Dipole-quadrupole potential · Quadrupole-quadrupole energy shift · Dipole-octupole interaction energy · Electric dipole-magnetic dipole potential · Diamagnetic contribution · Discriminatory energy shift

4.1 Introduction

While the leading electric dipole term of the multipolar series expansion Eq. (2.87) provides the dominant contribution to the van der Waals dispersion energy, as shown in the last Chapter, where the molecular QED calculation of the Casimir-Polder potential was presented, culminating in Eq. (3.6), there are plenty of

© The Author(s) 2016   57
A. Salam, *Non-Relativistic QED Theory of the van der Waals
Dispersion Interaction*, SpringerBriefs in Electrical and Magnetic Properties
of Atoms, Molecules, and Clusters, DOI 10.1007/978-3-319-45606-5_4

situations in which it is necessary to go beyond the electric dipole approximation, and consider higher-order terms. This is always the case if the atom or molecule sees a spatially non-uniform electromagnetic field. Another is if the size of the chromophore is comparable to or larger than the wavelength of radiation interacting with matter. Both of these situations may be dealt with by accounting for the possible variations at neighbouring points in space of the vector potential, $\vec{a}(\vec{r})$, whose mode expansion $\vec{a}(\vec{r})$ is therefore proportional to $e^{i\vec{k}\cdot\vec{r}} \approx 1 + i\vec{k}\cdot\vec{r} + \frac{(i\vec{k}\cdot\vec{r})^2}{2!} + \ldots$, where $\vec{k}$ is the wave vector. Truncating the expansion to a specific power of $\vec{r}$ enables a systematic correction to the leading electric dipole coupling term to be made in each of the electric polarisation, magnetisation and diamagnetisation fields, as shown in Chap. 2 when transforming from minimal-coupling to multipolar Hamiltonians. Since the interaction Hamiltonian is additive, particular multipole moment terms contributing to an application are easily selected at the start of a calculation.

For many previously studied systems, it is the first few electric terms of the multipolar series that are of interest. A classic example is provided by the helium dimer, detected spectroscopically in the early 1990s [1], and which is known to be very weakly bound, possessing a very shallow potential energy minimum, and which is described by a wave function that extends over large inter-nuclear separation distances. Including electric quadrupole and electric octupole coupling terms, as well as the effects of retardation on the dispersion potential, was necessary to ensure closer agreement with experiment was obtained. A similar feature was required to be included in the computation of interaction energies between lithium atoms when forming a dimer [2].

Other systems where higher multipoles are essential for an accurate description include dispersion forces between chiral molecules. Because such species possess low or no symmetry elements, spectroscopic selection rules are less restrictive, allowing contributions from electric quadrupole and magnetic dipole couplings to feature; in some cases transitions to all multipole orders is even possible. An interesting aspect of such interactions is that they are discriminatory, depending on the handedness of the coupled particles. Contributions to the dispersion energy arising from higher multipole moments are examined in this chapter.

4.2 Generalised Dispersion Energy Shift for Molecules with Arbitrary Electric Multipoles

In Chap. 3, the dominant contribution to the van der Waals dispersion potential arising between two neutral species was evaluated in the electric dipole approximation, producing the Casimir-Polder interaction energy of Eq. (3.6). Even though the magnetic dipole moment is of the same order of magnitude as the electric quadrupole moment, higher-order corrections to the electric polarisation field are commonly treated before magnetic effects are examined. In this section we generalise the energy shift between two electric dipole polarisable atoms or molecules to account for interactions between two arbitrarily electrically polarisable particles. Specific contributions to the dispersion energy involving electric quadrupole and octupole terms will be extracted in subsequent sections before going on to calculate potentials that include magnetic dipole and diamagnetic coupling terms.

Consider two polarisable species A and B positioned at $\vec{R}_A$ and $\vec{R}_B$, respectively, with $\vec{R} = \vec{R}_B - \vec{R}_A$ as before. In the multipolar coupling scheme, the interaction of an electric multipole moment of order l, located at $\vec{r}_\xi$ for particle ξ, given in reducible form as

$$E^{(l)}_{i_1 i_2 \ldots i_l}(\xi) = -\frac{e}{l!} (\vec{q}(\xi) - \vec{r}_\xi)_{i_1} (\vec{q}(\xi) - \vec{r}_\xi)_{i_2} \ldots (\vec{q}(\xi) - \vec{r}_\xi)_{i_l}, \qquad (4.1)$$

is expressed as

$$H^{(l)}_{int}(\xi) = -\varepsilon_0^{-1} E^{(l)}_{i_1 i_2 \ldots i_l}(\xi) \nabla_{i_2} \ldots \nabla_{i_l} d^{\perp}_{i_1}(\vec{r}_\xi). \qquad (4.2)$$

Because the divergence of the electric displacement field vanishes for a neutral source, $\vec{d}(\vec{r})$ is purely transverse. A consequence is that the coupling Hamiltonian Eq. (4.2) vanishes when the Cartesian tensor component labelling the field i_1 coincides with the index associated with any of the gradient operators that come before it.

Utilising the interaction Hamiltonian Eq. (4.2) instead of its electric dipole counterpart Eq. (3.14) in the fourth-order formula for the energy shift Eq. (3.15), along with summation over the twelve time-ordered diagrams featuring in the Casimir-Polder potential, in which two virtual photons of mode $(\vec{p}, \varepsilon)$ and $(\vec{p}', \varepsilon')$ are exchanged, produces for the dispersion interaction energy between a species A that is pure electrically polarisable with electric multipole moment of one kind of order a, and particle B that is similarly arbitrarily electrically polarisable with electric multipole moment of one type of order b, the intermediate expression

$$\Delta E^{(a,b)} = -\sum_{\vec{p},\vec{p}'} \sum_{r,s} \left(\frac{\hbar c p}{2\varepsilon_0 V}\right) \left(\frac{\hbar c p'}{2\varepsilon_0 V}\right) (\delta_{i_1 k_1} - \hat{p}_{i_1} \hat{p}_{k_1})(\delta_{j_1 l_1} - \hat{p}'_{j_1} \hat{p}'_{l_1}) [E^{(a)}_{i_1 \ldots i_a}(A)]^{0r} [E^{(a)}_{j_1 \ldots j_a}(A)]^{r0}$$

$$\times [E^{(b)}_{k_1 \ldots k_b}(B)]^{0s} [E^{(b)}_{l_1 \ldots l_b}(B)]^{s0} (\nabla_{i_2} \ldots \nabla_{i_a} \nabla_{k_2} \ldots \nabla_{k_b} e^{i\vec{p}\cdot\vec{R}})(\nabla_{j_2} \ldots \nabla_{j_a} \nabla_{l_2} \ldots \nabla_{l_b} e^{i\vec{p}'\cdot\vec{R}}) \sum_{x=1}^{12} D_x^{-1},$$

$$(4.3)$$

where the last sum is over the twelve energy denominator products, D_x^{-1}. The labels r and s denote intermediate electronic levels of A and B. The matrix element over states $|m>$ and $|n>$ of the electric multipole moment of order l, Eq. (4.1), is written as

$$<m|E^{(l)}_{i_1 \ldots i_l}(\xi)|n> = [E^{(l)}_{i_1 \ldots i_l}(\xi)]^{mn}. \qquad (4.4)$$

To facilitate the evaluation of the sums over $\vec{p}$ and $\vec{p}'$, the action of the gradient operators is made distinct by distinguishing between position vectors $\vec{R}$ and $\vec{R}'$ in the exponential factors, remembering to set them equal to one another at the end of the calculation. After performing the angular averages, Eq. (4.3) becomes

$$\Delta E^{(a,b)} = -\left(\frac{\hbar c}{2\varepsilon_0}\right)^2 \frac{1}{(2\pi)^4} \sum_{r,s} [E^{(a)}_{i_1...i_a}(A)]^{0r}[E^{(a)}_{j_1...j_a}(A)]^{r0}[E^{(b)}_{k_1...k_b}(B)]^{0s}[E^{(b)}_{l_1...l_b}(B)]^{s0}$$

$$\times (-\nabla^2\delta_{i_1 k_1} + \nabla_{i_1}\nabla_{k_1})^R(\nabla_{i_2}...\nabla_{i_a}\nabla_{k_2}...\nabla_{k_b})^R(-\nabla^2\delta_{j_1 l_1} + \nabla_{j_1}\nabla_{l_1})^{R'}(\nabla_{j_2}...\nabla_{j_a}\nabla_{l_2}...\nabla_{l_b})^{R'}\frac{1}{RR'}$$

$$\times \int_0^\infty\int_0^\infty \frac{1}{2}(\sin pR \sin p'R' + \sin pR' \sin p'R)\sum_{x=1}^{12} D_x^{-1}dpdp'|_{R=R'}.$$

$$(4.5)$$

Taking advantage of the fact that the sum of energy denominators is identical to that occurring in the Casimir-Polder calculation, the integral in Eq. (4.5) is then given by

$$\int_0^\infty\int_0^\infty \frac{1}{2}(\sin pR \sin p'R' + \sin pR' \sin p'R)\sum_{x=1}^{12} D_x^{-1}dpdp'$$

$$= \frac{2\pi k_{r0}k_{s0}}{(\hbar c)^3}\int_0^\infty du\frac{e^{-u(R+R')}}{(k_{r0}^2+u^2)(k_{s0}^2+u^2)}. \qquad (4.6)$$

Inserting Eq. (4.6) into (4.5) yields a generalised formula [3] for the pair dispersion potential between two arbitrarily electrically polarisable species, one of multipole order a, the other of order b,

$$\Delta E^{(a,b)} = -\frac{\hbar c}{32\pi^3\varepsilon_0^2}\int_0^\infty du\alpha^{(a)}_{i_1...i_a j_1...j_a}(A;iu)\alpha^{(b)}_{k_1...k_b l_1...l_b}(B;iu)$$

$$\times [(-\nabla^2\delta_{i_1 k_1} + \nabla_{i_1}\nabla_{k_1})(\nabla_{i_2}...\nabla_{i_a}\nabla_{k_2}...\nabla_{k_b})\frac{e^{-uR}}{R}]$$

$$\times [(-\nabla^2\delta_{j_1 l_1} + \nabla_{j_1}\nabla_{l_1})(\nabla_{j_2}...\nabla_{j_a}\nabla_{l_2}...\nabla_{l_b})\frac{e^{-uR}}{R}].$$

$$(4.7)$$

Result (4.7) holds for all separation distances outside contact of the orbitals describing the charge distribution of A and B for both species in fixed relative orientation to one another. The generalised anisotropic pure electric multipole moment dynamic polarisability of order l evaluated at the imaginary wave vector $k = iu$ that appears in formula (4.7) is defined as

$$\alpha^{(l)}_{i_1...i_l j_1...j_l}(\xi;iu) = \sum_t 2E_{t0}\frac{[E^{(l)}_{i_1...i_l}(\xi)]^{0t}[E^{(l)}_{j_1...j_l}(\xi)]^{t0}}{E_{t0}^2 + (\hbar cu)^2}. \qquad (4.8)$$

The Casimir-Polder potential for two oriented molecules is easily obtained from Eq. (4.7) on choosing $a = b = 1$, which corresponds to the electric dipole contribution. Whence

$$\Delta E^{(1,1)} = -\frac{\hbar c}{32\pi^3 \varepsilon_0^2} \int\limits_0^\infty du \, \alpha_{ij}^{(1)}(A; iu) \alpha_{kl}^{(1)}(B; iu) [(-\nabla^2 \delta_{ik} + \nabla_i \nabla_k) \frac{e^{-uR}}{R}][(-\nabla^2 \delta_{jl} + \nabla_j \nabla_l) \frac{e^{-uR}}{R}],$$

(4.9)

which is identical to Eq. (3.26) after substituting for $\alpha_{ij}^{(1)}(\xi; iu)$ from Eq. (4.8) and evaluating the gradients in Eq. (4.9). Expression (4.7) is now used to compute higher-order electric contributions to the two-body dispersive energy shift.

4.3   Electric Dipole-Quadrupole Dispersion Potential

The first higher-order energy shift to be considered is that between an electric dipole polarisable molecule, A, and an electric quadrupole polarisable one, B. For anisotropic particles, the interaction energy is obtained straightforwardly from general formula (4.7) on selecting $a = 1$ and $b = 2$, since from (4.1) these two values give rise to electric dipole and quadrupole moment operators, respectively. Thus

$$\Delta E^{(1,2)} = -\frac{\hbar c}{32\pi^3 \varepsilon_0^2} \int\limits_0^\infty du \, \alpha_{ij}^{(1)}(A; iu) \alpha_{klmn}^{(2)}(B; iu) H_{ikl}(uR) H_{jmn}(uR).$$

(4.10)

$H_{ijk}(uR)$ symbolises the geometric tensor

$$
\begin{aligned}
H_{ijk}(uR) &\equiv (-\nabla^2 \delta_{ij} + \nabla_i \nabla_j) \nabla_k \frac{e^{-uR}}{R} \\
&= [(\delta_{ij}\hat{R}_k + \delta_{jk}\hat{R}_i + \delta_{ik}\hat{R}_j - 5\hat{R}_i\hat{R}_j\hat{R}_k)(3 + 3uR + u^2 R^2) \\
&\quad + (\delta_{ij} - \hat{R}_i\hat{R}_j)\hat{R}_k(u^2 R^2 + u^3 R^3)]\frac{e^{-uR}}{R^4}.
\end{aligned}
$$

(4.11)

From Eq. (4.8), the electric quadrupole polarisability tensor at imaginary frequency is

$$\alpha_{ijkl}^{(2)}(\xi; iu) = \sum_t 2E_{t0} \frac{[E_{ij}^{(2)}(\xi)]^{0t}[E_{kl}^{(2)}(\xi)]^{t0}}{E_{t0}^2 + (\hbar cu)^2},$$

(4.12)

where the reducible form of the electric quadrupole operator located at $\vec{R}_\xi$ is, from Eq. (4.1),

$$E_{ij}^{(2)}(\xi) = -\frac{e}{2!}(\vec{q} - \vec{R}_\xi)_i(\vec{q} - \vec{R}_\xi)_j. \qquad (4.13)$$

As for the anisotropic Casimir-Polder potential, the energy shift for a randomly oriented pair of particles is obtained after carrying out separate orientational averages on each species. For the electric quadrupole polarisability Eq. (4.12), we use the fourth-rank Cartesian tensor average result [4]

$$<\alpha_{ijkl}^{(2)}(\xi)> = \frac{1}{10}(\delta_{ik}\delta_{jl} + \delta_{il}\delta_{jk})\alpha_{\lambda\mu\lambda\mu}^{(2)}(\xi), \qquad (4.14)$$

where the angular brackets denote the orientationally averaged quantity, and Greek subscripts designate Cartesian tensor components in the molecule-fixed axis. In arriving at Eq. (4.14), use has been made of the fact that $E_{ii}^{(2)}(\xi) = 0$, and $E_{ij}^{(2)}(\xi) = E_{ji}^{(2)}(\xi)$. The former property follows from the form of the electric quadrupole coupling,

$$H_{int}^{(2)}(\xi) = -\varepsilon_0^{-1}E_{ij}^{(2)}(\xi)\nabla_j d_i^{\perp}(\vec{R}_\xi), \qquad (4.15)$$

which vanishes when $i = j$, so that the trace of the quadrupole moment does not contribute to the interaction. The index symmetry relation follows immediately from Eq. (4.13). Therefore the electric quadrupole moment that features in this and other interactions is traceless, symmetric and of weight-2. It has irreducible components

$$E_{ij}^{(2)} = -\frac{e}{2!}[(\vec{q} - \vec{R}_\xi)_i(\vec{q} - \vec{R}_\xi)_j - \frac{1}{3}(\vec{q} - \vec{R}_\xi)^2\delta_{ij}]. \qquad (4.16)$$

Multiplying Eq. (4.14) with the tensor coming from the average of the electric dipole polarisability, $<\alpha_{ij}^{(1)}(\xi)> = \frac{1}{3}\delta_{ij}\alpha_{\lambda\lambda}^{(1)}(\xi)$, and contracting with the product of $H_{ijk}(uR)$ factors in Eq. (4.10) yields the isotropic energy shift [3]

$$\Delta E^{(1,2)} = -\frac{\hbar c}{160\pi^3\varepsilon_0^2 R^8} \int_0^\infty du\alpha^{(1)}(A; iu)\alpha_{\lambda\mu\lambda\mu}^{(2)}(B; iu)e^{-2uR}$$

$$\times [u^6R^6 + 6u^5R^5 + 27u^4R^4 + 84u^3R^3 + 162u^2R^2 + 180uR + 90].$$

$$(4.17)$$

Expression (4.17) has also been obtained using response theory [5–7], and by directly computing the two-photon scattering amplitude [8]. Asymptotic limits follow from Eq. (4.17) on making the usual approximations.

In the near-zone, an R^{-8} dependent interaction energy ensues,

$$\Delta E_{NZ}^{(1,2)} = -\frac{3}{8\pi^2\varepsilon_0^2 R^8}\sum_{r,s}\frac{|[\vec{E}^{(1)}(A)]^{0r}|^2[E_{\lambda\mu}^{(2)}(B)]^{0s}[E_{\lambda\mu}^{(2)}(B)]^{s0}}{E_{r0}+E_{s0}}, \tag{4.18}$$

which would be the result obtained on using a static electric dipole-quadrupole coupling potential of the form

$$V_{ijk}(\vec{R}) = \frac{3}{4\pi\varepsilon_0 R^4}E_i^{(1)}(A)E_{jk}^{(2)}(B)[(\delta_{ij}\hat{R}_k + \delta_{jk}\hat{R}_i + \delta_{ik}\hat{R}_j - 5\hat{R}_i\hat{R}_j\hat{R}_k)]. \tag{4.19}$$

This is evident on inspecting the u-independent term of $H_{ijk}(uR)$, Eq. (4.11), which has identical distance and orientational dependence to expression (4.19).

At very long-range the potential displays inverse ninth power law behaviour and is proportional to the static polarisabilities of each entity. It is given by the limiting form

$$\Delta E_{FZ}^{(1,2)} = -\frac{1593\hbar c}{1280\pi^3\varepsilon_0^2 R^9}\alpha^{(1)}(A;0)\alpha_{\lambda\mu\lambda\mu}^{(2)}(B;0). \tag{4.20}$$

4.4 Electric Quadrupole-Quadrupole Interaction Energy

The next higher-order contribution to the dispersion potential between particles with pure electric multipole polarisability characteristics is the energy shift between two electric quadrupole polarisable species. For oriented objects the interaction energy is readily obtained on inserting $a = b = 2$ into Eq. (4.7). This produces [3]

$$\Delta E^{(2,2)} = -\frac{\hbar c}{32\pi^3\varepsilon_0^2}\int_0^\infty du\alpha_{ijkl}^{(2)}(A;iu)\alpha_{pqrs}^{(2)}(B;iu)L_{ipjq}(uR)L_{krls}(uR). \tag{4.21}$$

The pure electric quadrupole polarisability was defined in Eq. (4.12). Effecting one more gradient on $H_{ijk}(uR)$ in Eq. (4.11) defines the geometric tensor $L_{ijkl}(uR)$ appearing in Eq. (4.21). It is given by

$$\begin{aligned}
L_{ijkl}(uR) &\equiv (-\nabla^2\delta_{ij} + \nabla_i\nabla_j)\nabla_k\nabla_l\frac{e^{-uR}}{R}\\
&= \{[(\delta_{ij}\delta_{kl} + \delta_{ik}\delta_{jl} + \delta_{jk}\delta_{il}) - 5(\delta_{ij}\hat{R}_k\hat{R}_l + \delta_{ik}\hat{R}_j\hat{R}_l + \delta_{il}\hat{R}_j\hat{R}_k + \delta_{jk}\hat{R}_i\hat{R}_l + \delta_{jl}\hat{R}_i\hat{R}_k + \delta_{kl}\hat{R}_i\hat{R}_j)\\
&\quad + 35\hat{R}_i\hat{R}_j\hat{R}_k\hat{R}_l)](3 + 3uR + u^2R^2)\\
&\quad + [\delta_{ij}(\delta_{kl} - 3\hat{R}_k\hat{R}_l) - (\delta_{ij}\hat{R}_k\hat{R}_l + \delta_{ik}\hat{R}_j\hat{R}_l + \delta_{il}\hat{R}_j\hat{R}_k + \delta_{jk}\hat{R}_i\hat{R}_l + \delta_{jl}\hat{R}_i\hat{R}_k + \delta_{kl}\hat{R}_i\hat{R}_j)\\
&\quad + 10\hat{R}_i\hat{R}_j\hat{R}_k\hat{R}_l)(u^2R^2 + u^3R^3) - (\delta_{ij} - \hat{R}_i\hat{R}_j)\hat{R}_k\hat{R}_l(u^4R^4)]\}\frac{e^{-uR}}{R^5}.
\end{aligned}$$

$$\tag{4.22}$$

To obtain the energy shift for isotropic A and B, we multiply the two L_{ijkl} factors and contract with tensors from the expression for averaged quadrupole polarisability Eq. (4.14). This results in the potential [3]

$$\Delta E^{(2,2)} = -\frac{\hbar c}{1600\pi^3 \varepsilon_0^2 R^{10}} \int_0^\infty du \, \alpha^{(2)}_{\lambda\mu\lambda\mu}(A; iu)\alpha^{(2)}_{\nu\pi\nu\pi}(B; iu)e^{-2uR}$$

$$\times [u^8 R^8 + 10u^7 R^7 + 89u^6 R^6 + 510u^5 R^5 + 1983u^4 R^4 + 5280u^3 R^3 + 9360u^2 R^2 + 10080uR + 5040].$$
$$(4.23)$$

At short separation distances expression (4.23) reduces to the near-zone limiting form

$$\Delta E^{(2,2)}_{NZ} = -\frac{63\hbar c}{20\pi^3 \varepsilon_0^2 R^{10}} \int_0^\infty du \, \alpha^{(2)}_{\lambda\mu\lambda\mu}(A; iu)\alpha^{(2)}_{\nu\pi\nu\pi}(B; iu), \qquad (4.24)$$

displaying inverse tenth power law dependence. It may be obtained directly on employing a static electric quadrupole-quadrupole coupling potential in second-order of perturbation theory. The form of the static quadrupolar coupling is given by the u-independent part of $L_{ijkl}(uR)$, Eq. (4.22). Retardation again weakens the far-zone potential by an additional factor of R^{-1}. From Eq. (4.23) the long-range asymptote is

$$\Delta E^{(2,2)}_{FZ} = -\frac{50319\hbar c}{6400\pi^3 \varepsilon_0^2 R^{11}} \alpha^{(2)}_{\lambda\mu\lambda\mu}(A; 0)\alpha^{(2)}_{\nu\pi\nu\pi}(B; 0), \qquad (4.25)$$

which is written in terms of static electric quadrupole polarisabilities.

4.5 Electric Dipole-Octupole Energy Shift

An interaction energy that is of the same order of magnitude as $\Delta E^{(2,2)}$ computed in the last section, occurs between an electric dipole polarisable species, chosen to be A, and an electric octupole polarisable molecule, B. For particles in fixed orientation relative to one another, the potential follows from the general formula Eq. (4.7) on choosing $a = 1$ and $b = 3$. Thus

$$\Delta E^{(1,3)} = -\frac{\hbar c}{32\pi^3 \varepsilon_0^2} \int_0^\infty du \, \alpha^{(1)}_{ij}(A; iu)\alpha^{(3)}_{klmrst}(B; iu)L_{iklm}(uR)L_{jrst}(uR), \qquad (4.26)$$

with $L_{ijkl}(uR)$ given by Eq. (4.22). From definition (4.8), the pure electric octupole polarisability tensor at imaginary frequency is

$$\alpha^{(3)}_{klmrst}(\xi; iu) = \sum_t 2E_{t0} \frac{[E^{(3)}_{klm}(\xi)]^{0t}[E^{(3)}_{rst}(\xi)]^{t0}}{E^2_{t0} + (\hbar cu)^2}. \tag{4.27}$$

The electric octupole moment appearing in Eq. (4.27) is in its reducible form,

$$E^{(3)}_{ijk}(\xi) = -\frac{e}{3!}(\vec{q} - \vec{R}_\xi)_i(\vec{q} - \vec{R}_\xi)_j(\vec{q} - \vec{R}_\xi)_k, \tag{4.28}$$

as obtained from Eq. (4.1). It is advantageous to decompose $E^{(3)}_{ijk}(\xi)$ into its irreducible components of weights-1 and -3 according to

$$E^{(3)}_{ijk}(\xi) = E^{(3,1)}_{ijk}(\xi) + E^{(3,3)}_{ijk}(\xi), \tag{4.29}$$

with the weight-1 term given by

$$E^{(3,1)}_{ijk}(\xi) = -\frac{e}{30}(\vec{q} - \vec{R}_\xi)^2[(\vec{q} - \vec{R}_\xi)_i\delta_{jk} + (\vec{q} - \vec{R}_\xi)_j\delta_{ik} + (\vec{q} - \vec{R}_\xi)_k\delta_{ij}], \tag{4.30}$$

and the weight-3 contribution given by

$$\begin{aligned}
E^{(3,3)}_{ijk}(\xi) = &-\frac{e}{3!}\{(\vec{q} - \vec{R}_\xi)_i(\vec{q} - \vec{R}_\xi)_j(\vec{q} - \vec{R}_\xi)_k \\
&- \frac{1}{5}(\vec{q} - \vec{R}_\xi)^2[(\vec{q} - \vec{R}_\xi)_i\delta_{jk} + (\vec{q} - \vec{R}_\xi)_j\delta_{ik} + (\vec{q} - \vec{R}_\xi)_k\delta_{ij}]\}.
\end{aligned} \tag{4.31}$$

The first of these has three-independent components and transforms like a vector, while Eq. (4.31) has seven independent components and features in both retarded and instantaneous interactions. From the form of the electric octupole coupling,

$$H^{(3)}_{int}(\xi) = -\varepsilon_0^{-1} E^{(3)}_{ijk}(\xi)\nabla_j\nabla_k d_i^{\perp}(\vec{R}_\xi), \tag{4.32}$$

and the decomposition Eq. (4.29), the following useful and easily verifiable relations follow:

$$E^{(3,3)}_{ijj}(\xi) = 0; \; E^{(3,1)}_{ijk}(\xi)E^{(3,3)}_{ijk}(\xi) = 0; \; E^{(3,1)}_{ijk}(\xi)E^{(3,1)}_{ijk}(\xi) = \frac{3}{5}E^{(3)}_{ijj}(\xi)E^{(3)}_{ikk}(\xi). \tag{4.33}$$

On rotational averaging, with electric octupole polarisability requiring use of the sixth-rank result [4], the isotropic energy shift may be separated into distinct octupole weight-1 and weight-3 dependent contributions according to [3]

$$\Delta E^{(1,3)} = -\frac{\hbar c}{1200\pi^3 \varepsilon_0^2 R^6} \int\limits_0^\infty du u^4 e^{-2uR} \alpha^{(1)}(A; iu) \alpha^{(3,1)}_{\lambda\mu\mu\lambda\nu\nu}(B; iu) [u^4 R^4 + 2u^3 R^3 + 5u^2 R^2 + 6uR + 3]$$

$$-\frac{\hbar c}{420\pi^3 \varepsilon_0^2 R^{10}} \int\limits_0^\infty du e^{-2uR} \alpha^{(1)}(A; iu) \alpha^{(3,3)}_{\lambda\mu\nu\lambda\mu\nu}(B; iu) [u^8 R^8 + 12u^7 R^7 + 90u^6 R^6 + 486u^5 R^5$$

$$+ 1863u^4 R^4 + 4950u^3 R^3 + 8775u^2 R^2 + 9450uR + 4725],$$

$$(4.34)$$

where $\alpha^{(3,1)}_{ijklmn}(\xi; iu)$ and $\alpha^{(3,3)}_{ijklmn}(\xi; iu)$ are electric octupole weight-1 and -3 imaginary frequency dependent polarisabilities, respectively. It is interesting to note that the first term of Eq. (4.34), proportional to $\alpha^{(3,1)}_{\lambda\mu\mu\lambda\nu\nu}(B; iu)$ may be understood as a higher-order correction to the Casimir-Polder energy shift. This is justified on account of selection rules for $E^{(3,1)}_{ijk}(\xi)$ being similar to those found in transitions involving $E^{(1)}_i(\xi)$, and that the coefficients of the polynomial term in powers of uR written within square brackets are identical to those occurring in the $\alpha^{(1)}(A; iu)\alpha^{(1)}(B; iu)$ potential.

Asymptotic limits of the electric dipole-electric octupole interaction follow from the potential applicable for all R, Eq. (4.34). In the near-zone we find the limit

$$\Delta E^{(1,3)}_{NZ} = -\frac{45\hbar c}{4\pi^3 \varepsilon_0^2 R^{10}} \int\limits_0^\infty du \alpha^{(1)}(A; iu) \alpha^{(3,3)}_{\lambda\mu\nu\lambda\mu\nu}(B; iu),$$

$$(4.35)$$

which depends only on octupole weight-3 components, and like $\Delta E^{(2,2)}_{NZ}$ exhibits an inverse tenth power law behaviour of R. At very long-range, Eq. (4.34) tends to

$$\Delta E^{(1,3)}_{FZ} = -\frac{\hbar c}{6720\pi^3 \varepsilon_0^2 R^{11}} \alpha^{(1)}(A; 0) \left[\frac{4473}{5} \alpha^{(3,1)}_{\lambda\mu\mu\lambda\nu\nu}(B; 0) + 190476 \alpha^{(3,3)}_{\lambda\mu\nu\lambda\mu\nu}(B; 0)\right], \quad (4.36)$$

which is proportional to R^{-11} and depends on both octupole weight-1 and -3 components.

4.6 Electric Dipole-Magnetic Dipole Potential

Thus far only electric multipole coupling has been considered and dispersion energies between atoms or molecules possessing pure electric multipole polarisabilities have been computed. Because the magnetic dipole moment, $\vec{m}(\xi)$, is of the same order of magnitude as the electric quadrupole moment, the former multipole moment should also be accounted for when evaluating the leading correction terms to the Casimir-Polder potential. In this section we present the calculation of the van der

Waals dispersion energy between an electric dipole polarisable particle, A, and a second species, B, that is magnetic dipole susceptible. We again employ the total system Hamiltonian for the two atoms or molecules and the radiation field Eq. (3.13), but now include the first and third terms of the interaction Hamiltonian written in multipolar theory, Eq. (2.87), giving the two-term coupling Hamiltonian

$$H_{int} = -\varepsilon_0^{-1}\vec{\mu}(A) \cdot \vec{d}^{\perp}(\vec{R}_A) - \vec{m}(B) \cdot \vec{b}(\vec{R}_B).\qquad(4.37)$$

The initial and final states remain the same as in all of the dispersion energy calculations considered, namely both species are in the ground electronic state with the electromagnetic field in the vacuum state. Exchange of two virtual photons between the pair mediates the interaction and the same twelve time-ordered diagrams encountered in the Casimir-Polder potential apply in the present case, with the electric dipole coupling vertices at centre B replaced by magnetic dipole ones. Fourth-order perturbation theory yields, after summing over virtual photon polarisations and momenta, the interaction energy for oriented A and B,

$$\Delta E = \frac{1}{8\pi^3\varepsilon_0^2\hbar c^3}\varepsilon_{jlq}\varepsilon_{ikp}\hat{R}_p\hat{R}_q \sum_{r,s}\mu_i^{0r}(A)\mu_j^{r0}(A)m_k^{0s}(B)m_l^{s0}(B)$$
$$\times \int_0^{\infty}\frac{duu^6e^{-2uR}k_{r0}k_{s0}}{(k_{r0}^2+u^2)(k_{s0}^2+u^2)}\left[\frac{1}{u^2R^2}+\frac{2}{u^3R^3}+\frac{1}{u^4R^4}\right].\qquad(4.38)$$

After orientational averaging, and defining the isotropic, real frequency dependent paramagnetic susceptibility of molecule ξ as

$$\chi_p(\xi;\omega) = \frac{2}{3}\sum_t\frac{|\vec{m}^{0t}(\xi)|^2E_{t0}^{\xi}}{(E_{t0}^{\xi})^2-(\hbar\omega)^2},\qquad(4.39)$$

energy shift (4.38) becomes [5–7, 9]

$$\Delta E = \frac{\hbar}{16\pi^3\varepsilon_0^2c}\int_0^{\infty}duu^6e^{-2uR}\alpha(A;iu)\chi_p(B;iu)\left[\frac{1}{u^2R^2}+\frac{2}{u^3R^3}+\frac{1}{u^4R^4}\right].\qquad(4.40)$$

The integral is over the complex variable $\omega = icu$, with the polarisability and susceptibility responding at imaginary frequency.

Asymptotically limiting forms of the potential follow from Eq. (4.40) on making the usual approximations. In the near-zone, $kR \ll 1$, resulting in

$$\Delta E_{NZ} = \frac{1}{72\pi^2\varepsilon_0^2\hbar^2c^4R^4}\sum_{r,s}|\vec{\mu}^{0r}(A)|^2|\vec{m}^{0s}(B)|^2E_{r0}E_{s0}(E_{r0}+E_{s0})^{-1},\qquad(4.41)$$

exhibiting an R^{-4} dependence on distance. Because there is no static coupling between an electric and a magnetic dipole, Eq. (4.41) does not represent a true near-zone limit. Energy shift (4.41) is in fact retarded, depending on the transition frequencies of each species. At very large R, Eq. (4.40) tends to the limit

$$\Delta E_{FZ} = \frac{7\hbar}{64\pi^3 \varepsilon_0^2 c R^7} \alpha(A;0)\chi_p(B;0). \tag{4.42}$$

The far-zone potential varies as R^{-7} and is proportional to the static molecular susceptibility functions. Interestingly, energy shift (4.40) and its limiting forms are positive in sign and are consequently repulsive.

The dispersion energy between two paramagnetic species is analogous in functional form to the Casimir-Polder potential Eq (3.6), and is obtained on replacing $\vec{\mu}^{0t}(\xi)$ by $\frac{1}{c}\vec{m}^{0t}(\xi)$. R^{-6} London and R^{-7} Casimir shifts of the form Eqs. (3.29) and (3.31) ensue on substituting transition electric dipole moments and polarisabilities by their magnetic dipole counterparts. Frequently, the far-zone potentials between two electric dipole polarisable atoms, two paramagnetically susceptible atoms, and one electric dipole polarisable species and one magnetic dipole polarisable atom, are combined as in

$$\Delta E_{FZ}^{em} = -\frac{\hbar}{64\pi^3 \varepsilon_0^2 R^7}\{23c[\alpha(A;0)\alpha(B;0) + \frac{1}{c^4}\chi_p(A;0)\chi_p(B;0)]$$
$$- \frac{7}{c}[\alpha(A;0)\chi_p(B;0) + \chi_p(A;0)\alpha(B;0)]\}, \tag{4.43}$$

and is commonly known in the literature as the Feinberg-Sucher potential [10], where the very last term of Eq. (4.43) signifies the potential between paramagnetically susceptible A and electric dipole polarisable B.

It should also be mentioned that interaction energy (4.40) has been obtained via response theory [5–7, 11, 12] starting from the expression

$$\Delta E = -\frac{1}{2\varepsilon_0^2}\sum_{modes} \alpha(A;\omega)\vec{d}^{\perp 2}(B;\vec{R}_A) - \frac{1}{2}\sum_{modes} \chi_p(B;\omega)\vec{b}^2(A;\vec{R}_B), \tag{4.44}$$

written here for isotropic molecules, and taking the expectation value over the total system ground state. The first term of (4.44) represents the response of electric dipole polarisable body A to the *magnetic* dipole dependent electric displacement field of species B at the position of A, $\vec{R}_A$, while the second term corresponds to the response of paramagnetically susceptible particle B to the *electric* dipole dependent magnetic field due to A at $\vec{R}_B$. Identical results are of course obtained on interchanging the roles of A and B.

4.7 Inclusion of Diamagnetic Coupling

From the form of the second term of Eq. (4.44) it is seen that the interaction energy between electric and magnetic dipole polarisable systems is proportional to the square of the magnetic field. Inspection of the fourth term written in the multipolar interaction Hamiltonian Eq. (2.87) reveals another contribution that is quadratic in $\vec{b}(\vec{r})$, namely the diamagnetic coupling term. Hence this contribution should also be included as it is of the same order as the paramagnetic component $-\frac{1}{2}\chi_p(B;\omega)\vec{b}^2(A;\vec{R}_B)$. With A electric dipole polarisable as before, the interaction Hamiltonian is

$$H_{int} = -\varepsilon_0^{-1}\vec{\mu}(A) \cdot \vec{d}^\perp(\vec{R}_A) + \frac{e^2}{12m}<q^2(B)> \vec{b}^2(\vec{R}_B),\qquad(4.45)$$

for isotropic diamagnetic coupling. Since this interaction term is quadratic in the electronic charge, third-order perturbation theory is required to evaluate the two-photon exchange contribution to the dispersion potential, the interaction vertex signifying a two-photon event due to the quadratic dependence on the magnetic field. Three time-ordered diagrams have to be summed over in a perturbative computation. One of the diagrams is illustrated in Fig. 4.1. The energy shift is found to be [5–7, 13]

$$\Delta E = -\frac{e^2}{144\pi^3\varepsilon_0^2 mc^2}\sum_r|\vec{\mu}^{0r}(A)|^2 <q^2(B)>^{00}\int_0^\infty \frac{du\,u^6 e^{-2uR}k_{r0}}{(k_{r0}^2+u^2)}\left[\frac{1}{u^2R^2}+\frac{2}{u^3R^3}+\frac{1}{u^4R^4}\right],$$

(4.46)

where $<q^2(B)>^{00}$ is the ground state matrix element of the electronic position variable, q.

In the near-zone, Eq. (4.46) leads to an R^{-5} dependence according to

$$\Delta E_{NZ} = -\frac{e^2}{288\pi^3\varepsilon_0^2 mc^2 R^5}\sum_r|\vec{\mu}^{0r}(A)|^2 <q^2(B)>^{00}k_{r0}.\qquad(4.47)$$

Fig. 4.1 One of three time-orderings contributing to the dispersion interaction between an electric dipole polarisable particle A and a diamagnetically susceptible species B

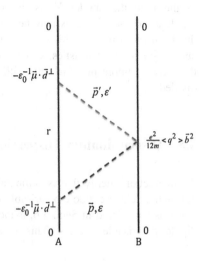

Meanwhile in the far-zone Eq. (4.46) reduces to the limit

$$\Delta E_{FZ} = -\frac{7e^2\hbar}{384\pi^3\varepsilon_0^2 mcR^7}\alpha(A;0) <q^2(B) > ^{00},\qquad(4.48)$$

which has inverse seventh power law form, and with the static electric dipole polarisability of A denoted by $\alpha(A;0)$.

If we define the magnetic susceptibility as the sum of the frequency dependent paramagnetic part given by Eq. (4.39), and a frequency independent diamagnetic contribution,

$$\chi_d(\xi;0) = -\frac{e^2}{6m}<q^2(\xi) > ^{00},\qquad(4.49)$$

according to

$$\chi(\xi;\omega) = \chi_p(\xi;\omega) + \chi_d(\xi;0),\qquad(4.50)$$

then the R^{-7} far-zone limits of the interaction of an electric dipole polarisable molecule with para- and dia-magnetic susceptible species Eqs. (4.42) and (4.48) may be combined to give

$$\Delta E_{FZ} = \frac{7\hbar}{64\pi^3\varepsilon_0^2 cR^7}\alpha(A;0)\chi(B;0).\qquad(4.51)$$

For molecules in the ground electronic state the static electric dipole polarisability and the static paramagnetic susceptibility are both greater than zero. $\chi(\xi;0)$ defined by Eq. (4.50), however, may be either positive or negative depending on the relative magnitudes of para- and dia-magnetic terms. If $\chi(\xi;0) < 0$, the molecule is classified as diamagnetic. The total interaction between two magnetically susceptible species may be written succinctly in terms of $\chi(\xi;\omega)$ using Eq. (4.50). Furthermore, the van der Waals dispersion potential between electric and magnetic dipole atoms or molecules may be expressed compactly on taking advantage of a duality transformation. This involves the simultaneous global exchange of electric and magnetic characteristics, which for the present case amounts to interchanging the electric dipole polarisability with the total magnetic susceptibility Eq. (4.50) divided by c^2.

4.8 Discriminatory Dispersion Potential

An interaction energy that is comparable in magnitude to the dispersion potential between an electric dipole polarisable body and a paramagnetically susceptible one that was considered in Sect. 4.6, is the energy shift between two electric-magnetic dipole polarisable species. This is the leading contribution to the dispersion potential involving coupling between two isotropic chiral molecules. Objects of this type possess few or no elements of symmetry, belonging to one of the following point groups: C_1, C_n, D_n, T, O and I, with $n \geq 2$. They lack both a centre of inversion and a plane of symmetry. Selection rules in such systems are consequently less restrictive, with higher-multipole moment contributions now allowed, often to all orders as in the case of an object with C_1 symmetry.

Retaining the dipole approximation, the interaction Hamiltonian Eq. (4.37) is modified to

$$H_{int} = -\varepsilon_0^{-1} \vec{\mu}(A) \cdot \vec{d}^{\perp}(\vec{R}_A) - \varepsilon_0^{-1} \vec{\mu}(B) \cdot \vec{d}^{\perp}(\vec{R}_B) - \vec{m}(A) \cdot \vec{b}(\vec{R}_A) - \vec{m}(B) \cdot \vec{b}(\vec{R}_B).$$

$$(4.52)$$

We are interested in extracting the interference term between electric dipole and magnetic dipole coupling terms. Also contained within Eq. (4.52) is the source of the Casimir-Polder potential, arising in the electric dipole approximation, its paramagnetic analogue proportional to $\chi_p(A; iu)\chi_p(B; iu)$, and the energy shift between pure electric dipole polarisable and pure magnetic dipole susceptible particles evaluated in Sect. 4.6. The computation of the discriminatory potential is similar to that carried out in Sect. 3.2 in which the electric dipole approximation was invoked. In the present case one electric dipole interaction vertex at each centre is replaced by a magnetic dipole coupling term. This results in a four-fold increase to forty-eight in the number of possible time-ordered sequences associated with two virtual photon exchange relative to the twelve diagrams featuring in the calculation of the Casimir-Polder potential. The twelve energy denominators are the same in both cases. Summing the graphs gives

$$\Delta E = -\sum_{\vec{p},\vec{p}'} \sum_{r,s} \left(\frac{\hbar p}{2\varepsilon_0 V}\right)\left(\frac{\hbar p'}{2\varepsilon_0 V}\right) \mu_i^{0r}(A) m_j^{r0}(A) \mu_k^{0s}(B) m_l^{s0}(B)[(\delta_{ik} - \hat{p}_i \hat{p}_k)(\delta_{jl} - \hat{p}_j' \hat{p}_l')$$

$$+ (\delta_{ik} - \hat{p}_i' \hat{p}_k')(\delta_{jl} - \hat{p}_j \hat{p}_l) + \varepsilon_{ilp}\varepsilon_{jkq}\hat{p}_p\hat{p}_q' + \varepsilon_{ilp}\varepsilon_{jkq}\hat{p}_q'\hat{p}_q]e^{i(\vec{p}+\vec{p}')\cdot\vec{R}} \sum_{x=1}^{12} D_x^{-1},$$

$$(4.53)$$

where ε_{ijk} is the skew-symmetric tensor, and the polarisation sums involving electric-magnetic and magnetic-magnetic vector combinations have been performed using the relations

$$\sum_{\varepsilon=1,2} e_i^{(\varepsilon)}(\vec{p})\bar{b}_j^{(\varepsilon)}(\vec{p}) = \varepsilon_{ijk}\hat{p}_k, \qquad (4.54)$$

and

$$\sum_{\varepsilon=1,2} b_i^{(\varepsilon)}(\vec{p})\bar{b}_j^{(\varepsilon)}(\vec{p}) = \delta_{ij} - \hat{p}_i \hat{p}_j. \qquad (4.55)$$

Next the wave vector sums are converted to integrals, the p' integral performed, and the variable p transformed to iu. In terms of the dyadics $\alpha_{ij} = \delta_{ij} - \hat{R}_i\hat{R}_j$ and $\beta_{ij} = \delta_{ij} - 3\hat{R}_i\hat{R}_j$, the analogue of Eq. (3.26) for two optically active molecules in fixed relative orientation is [7]

$$\Delta E = -\frac{1}{4\pi^3\varepsilon_0^2\hbar c^3}\sum_{r,s}\mu_i^{0r}(A)m_j^{r0}(A)\mu_k^{0s}(B)m_l^{s0}(B)\int_0^{\infty}\frac{duu^8e^{-2uR}}{(k_{r0}^2+u^2)(k_{s0}^2+u^2)}$$

$$\times\left[\begin{array}{l}(\alpha_{ik}\alpha_{jl}-\varepsilon_{ilp}\varepsilon_{jkq}\hat{R}_p\hat{R}_q)\dfrac{1}{u^2R^2}+(\alpha_{ik}\beta_{jl}+\beta_{ik}\alpha_{jl}-2\varepsilon_{ilp}\varepsilon_{jkq}\hat{R}_p\hat{R}_q)\dfrac{1}{u^3R^3}\\[2mm]+(\alpha_{ik}\beta_{jl}+\beta_{ik}\alpha_{jl}+\beta_{ik}\beta_{jl}-\varepsilon_{ilp}\varepsilon_{jkq}\hat{R}_p\hat{R}_q)\dfrac{1}{u^4R^4}+2\beta_{ik}\beta_{jl}\dfrac{1}{u^5R^5}+\beta_{ik}\beta_{jl}\dfrac{1}{u^6R^6}\end{array}\right].$$

$$(4.56)$$

Orientational averaging of (4.56) yields the dispersion potential for isotropic A and B,

$$\Delta E = -\frac{1}{18\pi^3\varepsilon_0^2\hbar c^3R^4}\sum_{r,s}[\vec{\mu}^{0r}(A)\cdot\vec{m}^{r0}(A)][\vec{\mu}^{0s}(B)\cdot\vec{m}^{s0}(B)]\int_0^{\infty}\frac{duu^4e^{-2uR}}{(k_{r0}^2+u^2)(k_{s0}^2+u^2)}\left[4+\frac{6}{uR}+\frac{3}{u^2R^2}\right],$$

$$(4.57)$$

which is valid for all R outside orbital overlap [7, 14]. Interestingly, the energy shifts (4.56) and (4.57) are discriminatory, depending on the handedness of each species. The chirality of A and B manifests through the pseudoscalar quantity $\vec{\mu}^{0r}(\xi)\cdot\vec{m}^{r0}(\xi)$, for randomly oriented chiral molecules, which changes sign when one enantiomer is replaced by its antipodal form. The chiral discrimination dispersion potential may also be expressed in terms of the mixed electric-magnetic dipole polarisability evaluated at imaginary frequency $\omega = icu$. At real frequency this quantity is defined as

$$G_{ij}(\xi;\omega) = \sum_t\left\{\frac{\mu_i^{0t}(\xi)m_j^{t0}(\xi)}{E_{t0}-\hbar\omega}+\frac{m_j^{0t}(\xi)\mu_i^{t0}(\xi)}{E_{t0}+\hbar\omega}\right\}.$$

$$(4.58)$$

For left- (L) and right- (R) handed isomers of the same chiral molecule, the L-L and R-R potential is repulsive, while the L-R and R-L energy shifts are attractive. For chemically distinct species the absolute sign of the interaction energy cannot be determined since the sign of the pseudoscalar $\vec{\mu}^{0t}(\xi)\cdot\vec{m}^{t0}(\xi)$ may be positive or negative. Changing either enantiomer from L→R or R→L will change the sign of ΔE.

For molecular separations R smaller than characteristic reduced transition wavelengths, the dominant term within square brackets of Eq. (4.57) is $3(uR)^{-2}$, the exponential factor in the same expression is approximated by unity, and the u-integral evaluated via Eq. (3.28) to give the near-zone asymptote

$$\Delta E_{NZ} = -\frac{1}{12\pi^2 \varepsilon_0^2 c^2 R^6} \sum_{r,s} \frac{[\vec{\mu}^{0r}(A) \cdot \vec{m}^{r0}(A)][\vec{\mu}^{0s}(B) \cdot \vec{m}^{s0}(B)]}{E_{r0} + E_{s0}}. \tag{4.59}$$

This result extends the London potential to dispersive inter-particle coupling between chiral species. It also exhibits R^{-6} behaviour. Expression (4.59) may also be obtained by employing static electric dipole-electric dipole and static magnetic dipole-magnetic dipole coupling potentials in second-order of perturbation theory, and extracting the cross-term.

In the far-zone limit $kR \gg 1$ and u^2 may be neglected in comparison to k_{r0} and k_{s0} in the energy denominators of Eq. (4.57). Performing the u-integral yields the asymptote

$$\Delta E_{FZ} = -\frac{\hbar^3 c}{3\pi^3 \varepsilon_0^2 R^9} \sum_{r,s} \frac{[\vec{\mu}^{0r}(A) \cdot \vec{m}^{r0}(A)][\vec{\mu}^{0s}(B) \cdot \vec{m}^{s0}(B)]}{E_{r0}^2 E_{s0}^2}, \tag{4.60}$$

having an R^{-9} dependence.

Finally, it should be pointed out that for oriented chiral molecules, there is a contribution to the discriminatory dispersion energy shift from two electric dipole-electric quadrupole polarisable species, and which is of the same order of magnitude as the $G(A; iu)G(B; iu)$ potential just presented. The mixed electric dipole-quadrupole polarisability tensor at real wave vector k is defined as

$$A_{ijk}(\xi; k) = \sum_t \left\{ \frac{\mu_i^{0t}(\xi)Q_{jk}^{t0}(\xi)}{E_{t0} - \hbar ck} + \frac{Q_{jk}^{0t}(\xi)\mu_i^{t0}(\xi)}{E_{t0} + \hbar ck} \right\}. \tag{4.61}$$

The discriminatory interaction energy proportional to $A_{ijk}(A; iu)A_{lmn}(B; iu)$ does not survive random orientational averaging, leaving formula (4.57) as the sole term contributing to the potential at this order of approximation for isotropic particles [7]. Response theory has also been successfully employed to compute interaction potentials between chiral systems [7, 12, 14, 15].

References

1. Luo F, Kim G, Giese CF, Gentry WR (1993) Influence of retardation on the higher-order multipole dispersion contributions to the helium dimer potential. J Chem Phys 99:10084
2. Yan Z-C, Dalgarno A, Babb JF (1997) Long-range interactions of lithium atoms. Phys Rev A 55:2882
3. Salam A, Thirunamachandran T (1996) A new generalisation of the Casimir-Polder potential to higher electric multipole polarisabilities. J Chem Phys 104:5094
4. Andrews DL, Thirunamachandran T (1977) On three-dimensional rotational averages. J Chem Phys 67:5026
5. Thirunamachandran T (1988) Vacuum fluctuations and intermolecular interactions. Phys Scr T21:123

6. Jenkins JK, Salam A, Thirunamachandran T (1994) Discriminatory dispersion interactions between chiral molecules. Mol Phys 82:835
7. Jenkins JK, Salam A, Thirunamachandran T (1994) Retarded dispersion interaction energies between chiral molecules. Phys Rev A 50:4767
8. Power EA, Thirunamachandran T (1996) Dispersion interactions between atoms involving electric quadrupole polarisabilities. Phys Rev A 53:1567
9. Mavroyannis C, Stephen MJ (1962) Dispersion forces. Mol Phys 5:629
10. Feinberg G, Sucher J (1968) General form of the retarded van der Waals potential. J Chem Phys 48:3333
11. Power EA, Thirunamachandran T (1993) Quantum electrodynamics with nonrelativistic sources. V. Electromagnetic field correlations and intermolecular interactions between molecules in either ground or excited states. Phys Rev A 47:2539
12. Salam A (1996) Intermolecular energy shifts between two chiral molecules in excited electronic states. Mol Phys 87:919
13. Salam A (2000) On the contribution of the diamagnetic coupling term to the two-body retarded dispersion interaction. J Phys B: At Mol Opt Phys 33:2181
14. Craig DP, Thirunamachandran T (1999) New approaches to chiral discrimination in coupling between molecules. Theo Chem Acc 102:112
15. Salam A (2010) Molecular quantum electrodynamics. Wiley, Hoboken

Chapter 5
van der Waals Dispersion Force Between Three Atoms or Molecules

Abstract Working within the electric dipole approximation, the leading non-additive retarded three-body dispersion potential is evaluated. Again two virtual photons are exchanged between each coupled pair. In order to avoid the use of sixth-order perturbation theory, necessitated when the interaction Hamiltonian is linear in the electric displacement field, a canonical transformation is performed on the $-\varepsilon_0^{-1}\vec{\mu} \cdot \vec{d}^{\perp}$ form to yield an effective two-photon coupling Hamiltonian that is quadratic in the displacement field. Third-order perturbation theory is then used to obtain the dispersion energy shift, which holds for a scalene triangle configuration. At short inter-particle separation distance, the Axilrod-Teller-Muto result originally derived via semi-classical theory is reproduced. Explicit expressions for interaction energies corresponding to equilateral and right-angled triangle, and collinear geometries are also presented, along with their asymptotically limiting forms. Whether the dispersion force is attractive or repulsive depends on the spatial arrangement of the three objects.

Keywords Two-photon coupling Hamiltonian · Retarded triple dipole potential · Axilrod-Teller-Muto interaction energy · Triangular and collinear geometries

5.1 Introduction

The dispersion energy is non-pairwise additive. Contributions therefore arise from three-, four-, ..., and many-body terms. It is important to note that the dominant contribution to the potential by far is due to pair interaction terms [1], led by the Casimir-Polder energy shift between two neutral non-polar species. There are a few chemical systems, however, for which retaining only the two-body term is insufficient. One example of longstanding and current interest is the interaction between elements in group 8 of the Periodic Table, whether they be in the gaseous or solid phases [2]. For the accurate computation of crystalline lattice energies, and

© The Author(s) 2016   75
A. Salam, *Non-Relativistic QED Theory of the van der Waals
Dispersion Interaction*, SpringerBriefs in Electrical and Magnetic Properties
of Atoms, Molecules, and Clusters, DOI 10.1007/978-3-319-45606-5_5

thermophysical properties such as the third virial coefficient, it is necessary to go beyond pairwise additivity and include the effects of three- and four-body terms in the dispersion potential. This has been done so as to calculate reliable interaction energies between three identical or different rare gas atoms [3], almost exclusively within the confines of semi-classical theory. Another example is the simulation and interpretation of photo- and magneto-association spectra [4]. Experimentally, these techniques are being employed to study the spectroscopy and dynamics of atoms and small molecules undergoing low energy collisions at ultracold temperatures. For instance, scattering cross-sections have been measured for atom-atom-atom and atom-diatom collisional processes involving alkali metal atoms and their homo- and hetero-nuclear dimer and trimer combinations in the study of the formation and dissociation of molecular species. Collisions at extremely low temperatures are especially sensitive to the long-range part of the interaction energy, and in an effort to improve the overall quality and transferability of such global potential energy surfaces, three body dispersion energy shifts have been accounted for [5], as they often make a significant contribution to the interaction energy. A similar motivation applies when considering interactions taking place in a medium, such as the influence of a solvent in modifying the strength of the force coupling embedded or constituent particles in any microscopic treatment of environmental effects.

The van der Waals dispersion potential between three electric dipole polarisable atoms is associated with the names of Axilrod and Teller, and Muto, who in the same year independently derived the nonpairwise additive three-body energy shift [6, 7]. They employed static dipolar coupling potentials, and so their expression for the interaction energy is valid only at short separation distances between particles. In keeping with the molecular QED presentation, the retardation corrected triple dipole dispersion interaction is evaluated in this Chapter. Perturbation theory calculations are again carried out. They are facilitated by employing an interaction Hamiltonian that is quadratic in the electric displacement field and is proportional to the polarisability of the atom or molecule [8]. It is shown how the Axilrod-Teller-Muto result follows as the near-zone asymptote of the potential applicable for all distances between the three objects. Since the sign of the interaction energy depends on the geometry adopted by the three atoms or molecules, explicit formulae for dispersion potentials for specific configurations are obtained, including an equilateral triangle arrangement, and three-bodies lying in a straight line.

5.2   Two-Photon Coupling: The Craig-Power Hamiltonian

Consider three mutually interacting neutral and polarisable atoms or molecules A, B and C, situated at $\vec{R}_A$, $\vec{R}_B$ and $\vec{R}_C$, respectively. The molecular QED Hamiltonian operator in the electric dipole approximation for this system is a straightforward extension of Eqs. (3.13) and (3.14) applicable to the two-body case,

$$H = H_{part}(A) + H_{part}(B) + H_{part}(C) + H_{rad}$$
$$- \varepsilon_0^{-1}\vec{\mu}(A) \cdot \vec{d}^{\perp}(\vec{R}_A) - \varepsilon_0^{-1}\vec{\mu}(B) \cdot \vec{d}^{\perp}(\vec{R}_B) - \varepsilon_0^{-1}\vec{\mu}(C) \cdot \vec{d}^{\perp}(\vec{R}_C). \qquad (5.1)$$

To arrive at the dispersion energy, the expectation value is taken over the ground state of the particles and field of the last three terms of Eq. (5.1), which comprise the perturbation operator. Hence the initial and final state specification of the total non-interacting system is given by

$$|0> \ = |0^A, 0^B, 0^C >, \qquad (5.2)$$

with 0^ξ, $\xi = A, B$ and C denoting that each species is in its lowest energy state. No photons are present in the electromagnetic field. As for pair dispersion potentials, coupling is viewed as due to the exchange of two virtual photons between any two interacting particles. Hence the use of the form of the perturbation operator dictated by Eq. (5.1), which is linear in the electric displacement field, would necessitate employing sixth-order perturbation theory for the evaluation of the three-body dispersion potential, and drawing and summing over 360 time-ordered sequences of photon absorption and emission events. These are associated with virtual photon of mode $(\vec{k}_1, \lambda_1)$ being exchanged between A and B, virtual photon of mode $(\vec{k}_2, \lambda_2)$ propagating between B and C, and C and A coupled via transfer of virtual photon of mode $(\vec{k}_3, \lambda_3)$, noting that the subscripts 1, 2 and 3 merely serve as labels to distinguish between the virtual photons, each polarisation and momentum ultimately being summed over.

Such a computation is technically demanding, labourious, and prone to error. At the end of Sect. 3.3 it was pointed out how the R^{-7} asymptotic form of the Casimir-Polder potential could be obtained using an interaction Hamiltonian that is quadratic in the electric displacement field. If the polarisability is taken to be frequency dependent, the ensuing effective coupling Hamiltonian may be used to evaluate the Casimir-Polder energy shift via diagrammatic perturbation theory or in a form of response theory, simplifying the computation in significant ways [9]. Below we show how the Craig-Power Hamiltonian [8] may be derived from Eq. (5.1), and use it in the next Section to compute the retarded triple dipole dispersion potential.

For a single centre at $\vec{R}$, the Hamiltonian in the electric dipole approximation is

$$H = H_{part} + H_{rad} - \varepsilon_0^{-1}\vec{\mu} \cdot \vec{d}^{\perp}(\vec{R}). \qquad (5.3)$$

To obtain the new Hamiltonian, a unitary transformation of the type Eq. (2.58) is carried out on Eq. (5.3), with the generating function S chosen such that the coupling term linear in the electric dipole moment and electric displacement field is removed for all energy non-conserving processes. With the use of identity Eq. (2.65), the new Hamiltonian correct to second-order is

$$\begin{aligned}
H_{new} &= H_{part} + H_{rad} - \varepsilon_0^{-1} \vec{\mu} \cdot \vec{d}^{\perp}(\vec{R}) + i[S, H_{part} + H_{rad}] \\
&\quad - i[S, -\varepsilon_0^{-1} \vec{\mu} \cdot \vec{d}^{\perp}(\vec{R})] + \frac{i}{2}[S, [S, H_{part} + H_{rad}]] + \cdots \\
&= H_{part} + H_{rad} - \frac{i}{2}[S, -\varepsilon_0^{-1} \vec{\mu} \cdot \vec{d}^{\perp}(\vec{R})] \\
&= H_{part} + H_{rad} + H_{int}^{new}.
\end{aligned} \tag{5.4}$$

The generator S is found from

$$i[S, H_{part} + H_{rad}] = \varepsilon_0^{-1} \vec{\mu} \cdot \vec{d}^{\perp}(\vec{R}). \tag{5.5}$$

Thus the new second-order Hamiltonian is $-\frac{i}{2}[S, H_{int}]$. Employing the particle and electromagnetic field base state $|E_n; N(\vec{k}, \lambda) >$, the off-diagonal matrix element for the particle factor, and the diagonal matrix element for the field factor of the generator is

$$<N(\vec{k}, \lambda); E_m | S | E_n; N(\vec{k}, \lambda) > = i \frac{\varepsilon_0^{-1} \vec{\mu}^{mn} \cdot \vec{d}^{\perp}(\vec{R})}{E_{mn} \pm \hbar c k}. \tag{5.6}$$

Substituting Eq. (5.6) into the new coupling Hamiltonian and evaluating the diagonal matrix element for particles results in the functional form

$$\begin{aligned}
<N(\vec{k}, \lambda); E_n | H_{int}^{new} | E_n; N(\vec{k}, \lambda) > &= -\frac{1}{2\varepsilon_0^2} \sum_m \frac{\vec{\mu}^{nm} \cdot \vec{d}^{\perp}(\vec{R}) \vec{\mu}^{mn} \cdot \vec{d}^{\perp}(\vec{R})}{E_{mn}^2 - (\hbar c k)^2} \\
&= -\frac{1}{2\varepsilon_0^2} <N(\vec{k}, \lambda); E_n | \alpha(k) \vec{d}^{\perp 2}(\vec{R}) | E_n; N(\vec{k}, \lambda) >,
\end{aligned} \tag{5.7}$$

where $\alpha(k)$ is the isotropic frequency-dependent electric dipole polarisability. The coupling Hamiltonian is now quadratic in the electric displacement field with the molecule responding through its polarisability [8, 10].

5.3 Triple Dipole Dispersion Potential

We now compute the leading non-additive contribution to the van der Waals dispersion potential between three electric dipole polarisable species described in the previous Section using the Craig-Power form of interaction operator. For oriented systems, this coupling, which is quadratic in the field, is given by the interaction Hamiltonian

$$H_{int} = -\frac{1}{2\varepsilon_0^2} \sum_{\text{modes}} \alpha_{ij}(A;k) d_i^\perp(\vec{R}_A) d_j^\perp(\vec{R}_A) - \frac{1}{2\varepsilon_0^2} \sum_{\text{modes}} \alpha_{kl}(B;k) d_k^\perp(\vec{R}_B) d_l^\perp(\vec{R}_B)$$

$$-\frac{1}{2\varepsilon_0^2} \sum_{\text{modes}} \alpha_{mn}(C;k) d_m^\perp(\vec{R}_C) d_n^\perp(\vec{R}_C).$$

$$(5.8)$$

A consequence of the quadratic dependence of the coupling Hamiltonian on the electric displacement field is that events leading to changes in the electromagnetic field involving two photons are readily dealt with. This is ideal for calculations of the dispersion potential in which particles are coupled through the exchange of two virtual photons.

For the interaction of three objects according to Eq. (5.8), with initial and final state specified by Eq. (5.2), the energy shift is evaluated using the third-order perturbation theory formula

$$\Delta E = \sum_{I,II} \frac{<0|H_{int}|II> <II|H_{int}|I> <I|H_{int}|0>}{E_{I0}E_{II0}},$$

$$(5.9)$$

where the sum is executed over all intermediate states $|I>$ and $|II>$, but not including $|0>$. Employing collapsed two-photon interaction vertices instead of coupling operators linear in $\vec{d}^\perp(\vec{r})$ means that there are only 3! time-ordered graphs to be summed over [11]. One of these is drawn in Fig. 5.1. The Arabic numerals 1, 2 and 3 in the diagram denote the mode of the virtual photon. The five remaining time-orderings are obtained on permuting the interaction vertices. Evaluation and summation produces

$$\Delta E = -\hbar c \sum_{\vec{k}_1} \sum_{\vec{k}_2} \sum_{\vec{k}_3} \left(\frac{k_1 k_2 k_3}{32\varepsilon_0^3 V^3} \right) [\alpha_{ij}(A;k_1) + \alpha_{ij}(A;k_3)][\alpha_{kl}(B;k_1) + \alpha_{kl}(B;k_2)]$$

$$\times [\alpha_{mn}(C;k_2) + \alpha_{mn}(C;k_3)](\delta_{ik} - \hat{k}_i^1 \hat{k}_k^1)(\delta_{lm} - \hat{k}_l^2 \hat{k}_m^2)(\delta_{jn} - \hat{k}_j^3 \hat{k}_n^3)e^{-i\vec{k}_1\cdot\vec{c}}e^{-i\vec{k}_2\cdot\vec{a}}e^{i\vec{k}_3\cdot\vec{b}}$$

$$\times \left[\frac{1}{(k_1+k_3)(k_2+k_3)} + \frac{1}{(k_1+k_2)(k_1+k_3)} + \frac{1}{(k_2+k_3)(k_1+k_2)} \right],$$

$$(5.10)$$

after performing the polarisation sums, and where $\hat{k}_i^x$, $x = 1, 2, 3$ denotes the ith Cartesian component of the wave vector of photon of mode x.

Expression (5.10) applies to a triangular arrangement of the three objects according to the relative displacement vectors $\vec{a} = \vec{R}_B - \vec{R}_C$, $\vec{b} = \vec{R}_C - \vec{R}_A$, and $\vec{c} = \vec{R}_A - \vec{R}_B$, which appear in the last equation. After converting the wave vector sums to integrals using relation (3.19), and performing the angular integrations, which are all of the type

Fig. 5.1 One of six diagrams
contributing to three-body
dispersion interaction when
two-photon coupling vertices
are employed

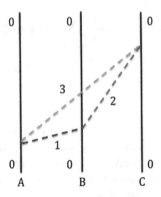

$$\frac{1}{4\pi}\int (\delta_{ij} - \hat{k}_i\hat{k}_j)e^{\pm i\vec{k}\cdot\vec{R}}d\Omega = \frac{1}{k^3}(-\nabla^2\delta_{ij} + \nabla_i\nabla_j)\frac{\sin kR}{R}, \qquad (5.11)$$

and expressing the sum of the three energy denominators as an integral over the parameter u according to [11]

$$\left[\frac{1}{(k_1+k_2)(k_2+k_3)} + \frac{1}{(k_2+k_3)(k_3+k_1)} + \frac{1}{(k_1+k_2)(k_1+k_3)}\right]$$
$$= \frac{4k_1k_2k_3}{\pi}\int_0^\infty du \frac{1}{(k_1^2+u^2)(k_2^2+u^2)(k_3^2+u^2)}, \qquad (5.12)$$

and transforming the wave vector integrals to the complex plane and evaluating using the result

$$\int_0^\infty dk\,\alpha_{ij}(\xi; k)\frac{k\sin kR}{k^2+u^2} = \frac{\pi}{2}\alpha_{ij}(\xi; iu)e^{-uR}, \qquad (5.13)$$

where $\alpha_{ij}(\xi; iu)$ is the polarisability tensor at imaginary frequency, yields the retarded triple dipole dispersion potential [11, 12] for oriented A, B and C

$$\Delta E = -\frac{\hbar c}{64\pi^4\varepsilon_0^3}(-\nabla^2\delta_{km} + \nabla_k\nabla_m)^a(-\nabla^2\delta_{in} + \nabla_i\nabla_n)^b(-\nabla^2\delta_{jl} + \nabla_j\nabla_l)^c\frac{1}{abc}$$
$$\times \int_0^\infty du\,\alpha_{ij}(A; iu)\alpha_{kl}(B; iu)\alpha_{mn}(C; iu)e^{-u(a+b+c)}.$$
$$(5.14)$$

Result (5.14) for rotationally averaged particles was first obtained by Aub and Zienau [13]. Their calculation was carried out using covariant S-matrix theory of quantum electrodynamics. From the sixth-order perturbation term of the scattering matrix, they were able to extract the non-relativistic limit to the potential proportional to the electric dipole polarisabilities of the three species. Differing physical viewpoints for the manifestation of the interaction within Coulomb gauge QED were subsequently used by a few workers [14–18] to confirm the three-body dispersion energy shift.

Performing an orientational average on Eq. (5.14), and using the definition of the $F_{ij}(pR)$ tensor, Eq. (3.23), which after inserting $p = iu$ in the latter and multiplying by u^3 is given by

$$F_{ij}(uR) \equiv (-\nabla^2 \delta_{ij} + \nabla_i \nabla_j)\frac{e^{-uR}}{R}$$

$$= -[(\delta_{ij} - \hat{R}_i\hat{R}_j)u^2R^2 + (\delta_{ij} - 3\hat{R}_i\hat{R}_j)(uR+1)]\frac{e^{-uR}}{R^3}, \qquad (5.15)$$

the retarded triple dipole dispersion potential may be written as

$$\Delta E = \frac{\hbar c}{64\pi^4 \varepsilon_0^3} \int_0^\infty du\, \alpha(A;iu)\alpha(B;iu)\alpha(C;iu)F_{jk}(ua)F_{ij}(ub)F_{ik}(uc). \qquad (5.16)$$

Substituting the second equality of Eq. (5.15) into (5.16) and evaluating the product of the three geometrical factors produces the lengthy expression [12]

$$\Delta E = \frac{\hbar c}{64\pi^4 \varepsilon_0^3}\frac{1}{a^3b^3c^3}\int_0^\infty du\, e^{-u(a+b+c)}\alpha(A;iu)\alpha(B;iu)\alpha(C;iu)$$

$$\times \{[-6+9[(\hat{a}\cdot\hat{b})^2 + (\hat{b}\cdot\hat{c})^2 + (\hat{c}\cdot\hat{a})^2] - 27(\hat{a}\cdot\hat{b})(\hat{b}\cdot\hat{c})(\hat{c}\cdot\hat{a})](1+ua)(1+ub)(1+uc)$$

$$+ [-4+3[3(\hat{a}\cdot\hat{b})^2 + (\hat{b}\cdot\hat{c})^2 + (\hat{c}\cdot\hat{a})^2] - 9(\hat{a}\cdot\hat{b})(\hat{b}\cdot\hat{c})(\hat{c}\cdot\hat{a})](1+ua)(1+ub)u^2c^2$$

$$+ [-4+3[(\hat{a}\cdot\hat{b})^2 + 3(\hat{b}\cdot\hat{c})^2 + (\hat{c}\cdot\hat{a})^2] - 9(\hat{a}\cdot\hat{b})(\hat{b}\cdot\hat{c})(\hat{c}\cdot\hat{a})]u^2a^2(1+ub)(1+uc)$$

$$+ [-4+3[(\hat{a}\cdot\hat{b})^2 + (\hat{b}\cdot\hat{c})^2 + 3(\hat{c}\cdot\hat{a})^2] - 9(\hat{a}\cdot\hat{b})(\hat{b}\cdot\hat{c})(\hat{c}\cdot\hat{a})](1+ua)u^2b^2(1+uc)$$

$$+ [-2+3(\hat{a}\cdot\hat{b})^2 + 3(\hat{b}\cdot\hat{c})^2 + (\hat{c}\cdot\hat{a})^2 - 3(\hat{a}\cdot\hat{b})(\hat{b}\cdot\hat{c})(\hat{c}\cdot\hat{a})]u^2a^2(1+ub)u^2c^2$$

$$+ [-2+3(\hat{a}\cdot\hat{b})^2 + (\hat{b}\cdot\hat{c})^2 + 3(\hat{c}\cdot\hat{a})^2 - 3(\hat{a}\cdot\hat{b})(\hat{b}\cdot\hat{c})(\hat{c}\cdot\hat{a})](1+ua)u^2b^2u^2c^2$$

$$+ [-2+(\hat{a}\cdot\hat{b})^2 + 3(\hat{b}\cdot\hat{c})^2 + 3(\hat{c}\cdot\hat{a})^2 - 3(\hat{a}\cdot\hat{b})(\hat{b}\cdot\hat{c})(\hat{c}\cdot\hat{a})]u^2a^2u^2b^2(1+uc)$$

$$+ [(\hat{a}\cdot\hat{b})^2 + (\hat{b}\cdot\hat{c})^2 + (\hat{c}\cdot\hat{a})^2 - (\hat{a}\cdot\hat{b})(\hat{b}\cdot\hat{c})(\hat{c}\cdot\hat{a})]u^2a^2u^2b^2u^2c^2\}.$$

$$(5.17)$$

Result (5.17) applies for all separation distances a, b and c outside the charge overlap region, extending out to infinity. Asymptotically limiting forms are obtained in the next Section.

5.4   Far- and Near-Zone Limits

Identical physical arguments and mathematical approximations invoked to obtain the short- and long-range asymptotes to the Casimir-Polder potential (see Sect. 3.3) may also be applied to the three-body dispersion potential Eq. (5.14). In the far-zone the relative displacements are large compared to characteristic reduced transition wavelengths, and low virtual photon frequency values contribute most to the energy shift in this regime on appeal to the time-energy uncertainty principle. Hence the polarisabilities appearing in Eq. (5.14) may be taken to be static and placed outside of the integral. The interaction is therefore governed entirely by the dynamics of the electromagnetic field, the particles being passive. The u-integral in Eq. (5.14) is then immediate, yielding

$$\Delta E_{FZ} = -\frac{\hbar c}{64\pi^4 \varepsilon_0^3} \alpha_{ij}(A;0)\alpha_{kl}(B;0)\alpha_{mn}(C;0)$$

$$\times \; (-\nabla^2 \delta_{km} + \nabla_k \nabla_m)^a (-\nabla^2 \delta_{in} + \nabla_i \nabla_n)^b (-\nabla^2 \delta_{jl} + \nabla_j \nabla_l)^c \frac{1}{abc(a+b+c)},$$

$$(5.18)$$

for oriented species forming a scalene triangle.

At the opposite extreme $a \ll k_{r0}^{-1}$, $b \ll k_{s0}^{-1}$ and $c \ll k_{t0}^{-1}$, where r, s and t designate intermediate electronic states of A, B and C, respectively. Now the dynamics of the particles are important, with coupling to the radiation field taken to be static. Hence the exponential term in Eq. (5.14) is approximated to unity, giving the near-zone limit

$$\Delta E_{NZ} = -\frac{\hbar c}{64\pi^4 \varepsilon_0^3} (-\nabla^2 \delta_{km} + \nabla_k \nabla_m)^a (-\nabla^2 \delta_{in} + \nabla_i \nabla_n)^b (-\nabla^2 \delta_{jl} + \nabla_j \nabla_l)^c \frac{1}{abc}$$

$$\times \int_0^\infty du \, \alpha_{ij}(A;iu)\alpha_{kl}(B;iu)\alpha_{mn}(C;iu),$$

$$(5.19)$$

which is written in an alternative form to that first derived by Axilrod-Teller, and by Muto. Noting that the evaluation of gradients produces

$$(-\nabla^2 \delta_{ij} + \nabla_i \nabla_j)\frac{1}{R} = (\delta_{ij} - 3\widehat{R}_i\widehat{R}_j)R^{-3}, \qquad (5.20)$$

and generalising the u-integral Eq. (3.28) to

$$rst \int_0^\infty \frac{du}{(r^2+u^2)(s^2+u^2)(t^2+u^2)} = \frac{\pi}{2}\frac{(r+s+t)}{(r+s)(s+t)(t+r)}, \qquad (5.21)$$

result (5.19) is more commonly expressed as

$$\Delta E_{NZ} = \frac{1}{432\pi^3 \varepsilon_0^3} \frac{1}{a^3 b^3 c^3} (\delta_{ij} - 3\hat{a}_i \hat{a}_j)(\delta_{jk} - 3\hat{b}_j \hat{b}_k)(\delta_{ik} - 3\hat{c}_i \hat{c}_k)$$

$$\times \sum_{r,s,t} |\vec{\mu}^{0r}(A)|^2 |\vec{\mu}^{0s}(B)|^2 |\vec{\mu}^{0t}(C)|^2 \frac{(E_{r0} + E_{s0} + E_{t0})}{(E_{r0} + E_{s0})(E_{s0} + E_{t0})(E_{t0} + E_{r0})},$$

(5.22)

on orientational averaging. Expression (5.22) is equivalent to the result obtained from Eq. (5.17) on letting $e^{-u(a+b+c)} \rightarrow 1$ and retaining the u-independent term, corresponding to the first geometrical term within braces, and on using Eqs. (5.20) and (5.21). Analogous to the London dispersion potential, the short-range triple dipole interaction energy may be derived directly by employing static dipolar coupling potentials in third-order of perturbation theory, as done originally in Refs. [6, 7].

Introducing the direction cosines $\cos\theta_A = -\hat{b} \cdot \hat{c}$, $\cos\theta_B = -\hat{c} \cdot \hat{a}$ and $\cos\theta_C = -\hat{a} \cdot \hat{b}$ where θ_A, θ_B and θ_C are internal angles of the triangle formed by A, B and C facing sides extended by BC, CA and AB, respectively, with lengths a, b and c, as shown in Fig. 5.2, the product of the orientational factors yields

$$\{-6 + 9[(\hat{a} \cdot \hat{b})^2 + (\hat{b} \cdot \hat{c})^2 + (\hat{c} \cdot \hat{a})^2] - 27(\hat{a} \cdot \hat{b})(\hat{b} \cdot \hat{c})(\hat{c} \cdot \hat{a})\}$$
$$= 3[1 + 3\cos\theta_A \cos\theta_B \cos\theta_C],$$

(5.23)

on using the identity

$$(\hat{a} \cdot \hat{b})^2 + (\hat{b} \cdot \hat{c})^2 + (\hat{c} \cdot \hat{a})^2 = 1 - 2\cos\theta_A \cos\theta_B \cos\theta_C.$$

(5.24)

Hence the near-zone potential may be written in more recognisable form as

$$\Delta E_{NZ} = \frac{1}{144\pi^3 \varepsilon_0^3} \sum_{r,s,t} |\vec{\mu}^{0r}(A)|^2 |\vec{\mu}^{0s}(B)|^2 |\vec{\mu}^{0t}(C)|^2 \frac{(E_{r0} + E_{s0} + E_{t0})}{(E_{r0} + E_{s0})(E_{s0} + E_{t0})(E_{t0} + E_{r0})}$$

$$\times \frac{[1 + 3\cos\theta_A \cos\theta_B \cos\theta_C]}{a^3 b^3 c^3}.$$

(5.25)

A cubic dependence on each separation distance is exhibited.

Fig. 5.2 Triangle configuration for the three bodies. Relative displacements and internal angles are defined in the text

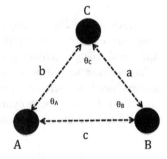

The potentials obtained in this and the previous Section are now applied to particular geometrical arrangements of the three bodies.

5.5 Equilateral Triangle Geometry

The first configuration to be considered is an equilateral triangle arrangement of the three objects, whose general disposition is shown in Fig. 5.2. We choose the lengths of each side to be equal, with $a = b = c = R$, and the interior angles $\theta_A = \theta_B = \theta_C = 60°$. Inserting these parameters into Eq. (5.17) produces [19]

$$
\Delta E = \frac{\hbar c}{512\pi^4 \varepsilon_0^3 R^9} \int_0^\infty du\, e^{-3uR} \alpha(A; iu)\alpha(B; iu)\alpha(C; iu)
$$

$$
\times\ [7u^6 R^6 + 3u^5 R^5 + 24u^4 R^4 + 75u^3 R^3 + 120u^2 R^2 + 99uR + 33],
$$

$$(5.26)$$

which holds for all values of R beyond the contact radius of the three particles. It is interesting to note that the energy shift is positive, indicating a repulsive potential for this geometry.

Near- and far-zone limiting potentials follow from Eq. (5.26) in the usual way. For short inter-object separations, $uR \ll 1$. Thus $e^{-3uR} \approx 1$, and the uR-independent term within square brackets is retained, giving

$$
\Delta E_{NZ} = \frac{33\hbar c}{512\pi^4 \varepsilon_0^3 R^9} \int_0^\infty du\, \alpha(A; iu)\alpha(B; iu)\alpha(C; iu),
$$

$$(5.27)$$

which has inverse ninth power law dependence. Substituting for the polarisabilities at imaginary wave vector and performing the u-integral via Eq. (5.21) yields the familiar form obtained by Axilrod-Teller and Muto explicitly in terms of transition electric dipole moments,

$$
\Delta E_{NZ} = \frac{11}{8} \frac{1}{144\pi^3 \varepsilon_0^3 R^9} \sum_{r,s,t} |\vec{\mu}^{0r}(A)|^2 |\vec{\mu}^{0s}(B)|^2 |\vec{\mu}^{0t}(C)|^2 \frac{(E_{r0} + E_{s0} + E_{t0})}{(E_{r0} + E_{s0})(E_{s0} + E_{t0})(E_{t0} + E_{r0})}.
$$

$$(5.28)$$

Expression (5.28) is identical to the result that would be obtained directly on using the near-zone asymptote Eq. (5.25) on selecting variables appropriate to equilateral triangle geometry, the distance and angular factors giving $11/8R^9$.

To evaluate the far-zone asymptote, in which $uR \gg 1$, the polarisabilities are approximated by their zero frequency forms, and taken outside of the integral in Eq. (5.26). Each of the polynomial terms is then integrated using Eq. (3.30) and summed to give the interaction energy at long-range

$$\Delta E_{FZ} = \frac{2^4 \times 79}{3^5} \frac{\hbar c}{64\pi^4 \varepsilon_0^3 R^{10}} \alpha(A;0)\alpha(B;0)\alpha(C;0), \tag{5.29}$$

in agreement with the result first obtained by Aub and Zienau [13], and subsequently confirmed by others [15, 17, 18]. Compared to the near-zone potential, the far-zone shift is weakened by an inverse power of R, displaying R^{-10} behaviour overall. Both asymptotes are repulsive.

5.6 Collinear Arrangement

We now examine the situation in which all three species lie along a line. One possible configuration is to place object C mid-way between particles A and B, displaced from each by a distance R. This corresponds to collapsing the equilateral triangle, resulting in a collinear arrangement. Hence $2a = 2b = c = R$, and direction cosines $\hat{a} \cdot \hat{b} = 1$, $\hat{b} \cdot \hat{c} = \hat{c} \cdot \hat{a} = -1$, for which $\theta_C = 180°$ and $\theta_A = \theta_B = 0°$. Inserting these values in the potential for arbitrary geometry Eq. (5.17) produces [19]

$$\Delta E = \frac{\hbar c}{8\pi^4 \varepsilon_0^3 R^9} \int_0^\infty du e^{-2uR} \alpha(A;iu)\alpha(B;iu)\alpha(C;iu)$$
$$\times [u^6 R^6 + 5u^5 R^5 + 17u^4 R^4 + 16u^3 R^3 - 36u^2 R^2 - 96uR - 48], \tag{5.30}$$

and which is applicable for all R.

After the usual approximations, the short-range limit is found to be

$$\Delta E_{NZ} = -\frac{6\hbar c}{\pi^4 \varepsilon_0^3 R^9} \int_0^\infty du \alpha(A;iu)\alpha(B;iu)\alpha(C;iu), \tag{5.31}$$

exhibiting attractive R^{-9} behaviour. Explicitly in terms of transition dipoles and energies, Eq. (5.31) may be written as

$$\Delta E_{NZ} = -\frac{8}{9}\frac{1}{\pi^3 \varepsilon_0^3 R^9} \sum_{r,s,t} |\vec{\mu}^{0r}(A)|^2 |\vec{\mu}^{0s}(B)|^2 |\vec{\mu}^{0t}(C)|^2 \frac{(E_{r0} + E_{s0} + E_{t0})}{(E_{r0} + E_{s0})(E_{s0} + E_{t0})(E_{t0} + E_{r0})}, \tag{5.32}$$

which is identical to the energy shift obtained from limit (5.25) on noting that the geometrical factor for this particular configuration from this last equation is $-2^7/R^9$.

On letting $\omega \to 0$ in the electric dipole polarisabilities appearing in Eq. (5.30) so that their static forms may be factored outside of the integral, performing the u-integral yields the far-zone asymptote

$$\Delta E_{FZ} = -\frac{93\hbar c}{32\pi^4 \varepsilon_0^3 R^{10}} \alpha(A;0)\alpha(B;0)\alpha(C;0),\qquad(5.33)$$

which is also attractive in nature, and like its equilateral triangle counterpart, displays R^{-10} dependence on distance [19].

5.7   Right-Angled Triangle Configuration

A third configuration to be considered is a right-angled triangle, with C vertically above A. For lengths $b = 3R$, $c = 4R$, a is then equal to $5R$, and the direction cosines for this 3-4-5 triangle are simply $\cos\theta_A = -\hat{b}\cdot\hat{c} = 0$, $\cos\theta_B = -\hat{c}\cdot\hat{a} = \frac{4}{5}$, and $\cos\theta_C = -\hat{a}\cdot\hat{b} = \frac{3}{5}$. Inserting these parameters in Eq. (5.17) yields the triple dipole dispersion interaction energy

$$\Delta E = \frac{\hbar c}{5400000 \times 64\pi^4 \varepsilon_0^3 R^9} \int_0^\infty due^{-12uR}\alpha(A;iu)\alpha(B;iu)\alpha(C;iu)$$

$$\times\ [90000u^6R^6 + 15900u^5R^5 + 146150u^4R^4 + 9588u^3R^3 + 4003u^2R^2 + 900uR + 75].$$
$$(5.34)$$

In the near-zone, retaining the u-independent term within square brackets and approximating e^{-12uR} by unity in Eq. (5.34) leads to

$$\Delta E_{NZ} = \frac{3\hbar c}{64\pi^4 \varepsilon_0^3 R^9}\frac{1}{3^3\times 4^3\times 5^3}\int_0^\infty du\alpha(A;iu)\alpha(B;iu)\alpha(C;iu),\qquad(5.35)$$

exhibiting inverse ninth power dependence on separation distance. This is identical to the result obtained directly from Eq. (5.25), the distance and angular factors reducing to $(3\times 4\times 5\times R^3)^{-3}$. In the far-zone Eq. (5.34) tends to the limiting form

$$\Delta E_{FZ} = \frac{4.05\times 10^{-4}\hbar c}{60\times 64\pi^4 \varepsilon_0^3 R^{10}} \alpha(A;0)\alpha(B;0)\alpha(C;0),\qquad(5.36)$$

displaying inverse tenth power law dependence.

Whether a three-body energy shift is attractive or repulsive depends on the configuration adopted. The asymptotic limits, however, may differ in sign from the general result, and from each other, that is, the near-zone potential may be attractive and the far-zone limit repulsive, or vice versa. This is due to the fact that the short-range limit is calculated by retaining the coefficient of the u-independent term in the polynomial factor, while at very long range, all terms have to be retained and integrated, and individual terms may be of either sign. Their sum then determines the overall sign of the interaction energy in the far-zone.

References

1. Margenau H, Kestner NR (1969) Theory of intermolecular forces. Pergamon, Oxford
2. Maitland GC, Rigby M, Smith EB, Wakeham WA (1981) Intermolecular forces. Clarendon, Oxford
3. Tang L-Y, Yan Z-C, Shi T-Y, Babb JF, Mitroy J (2012) The long-range non-additive three-body dispersion interactions for the rare gases, alkali, and alkaline Earth atoms. J Chem Phys 136:104104
4. Stwalley WC, Wang H (1999) Photoassociation of ultracold atoms: a new spectroscopic technique. J Mol Spectrosc 195:194
5. Cvitas MJ, Soldan P, Hutson JM (2006) Long-range intermolecular forces in triatomic systems: connecting the atom-diatom and atom-atom-atom representations. Mol Phys 104:23
6. Axilrod BM, Teller E (1943) Interaction of the van der Waals type between three atoms. J Chem Phys 11:299
7. Muto Y (1943) Force between nonpolar molecules. J Phys Math Soc Jpn 17:629
8. Craig DP, Power EA (1969) The asymptotic Casimir-Polder potential from second-order perturbation theory and its generalisation for anisotropic polarisabilities. Int J Quant Chem 3:903
9. Power EA, Thirunamachandran T (1993) Quantum electrodynamics with nonrelativistic sources. V. Electromagnetic field correlations and intermolecular interactions between molecules in either ground or excited states. Phys Rev A 47:2539
10. Power EA, Thirunamachandran T (1993) A new insight into the mechanism of intermolecular forces. Chem Phys 171:1
11. Passante R, Power EA, Thirunamachandran T (1998) Radiation-molecule coupling using dynamic polarisabilities: application to many-body forces. Phys Lett A 249:77
12. Salam A (2014) Dispersion potential between three bodies with arbitrary electric multipole polarisabilities: molecular QED theory. J Chem Phys 140:044111
13. Aub MR, Zienau S (1960) Studies on the retarded interaction between neutral atoms. I. Three-body London-van der Waals interaction of neutral atoms. Proc Roy Soc London A257:464
14. McLachlan AD (1963) Three-body dispersion forces. Mol Phys 6:423
15. Power EA, Thirunamachandran T (1985) The non-additive dispersion energies for N molecules: a quantum electrodynamical theory. Proc Roy Soc London A401:267
16. Power EA, Thirunamachandran T (1994) Zero-point energy differences and many-body dispersion forces. Phys Rev A 50:3929
17. Cirone M, Passante R (1996) Vacuum field correlations and the three-body Casimir-Polder potential. J Phys B: At Mol Opt Phys 29:1871
18. Thirunamachandran T (1988) Vacuum fluctuations and intermolecular interactions. Phys Scr T21:123
19. Salam A (2013) Higher-order electric multipole contributions to retarded non-additive three-body dispersion interaction energies between atoms: equilateral triangle and collinear configurations. J Chem Phys 139:244105

Chapter 6
Three-Body Dispersion Energy Shift: Contributions from Higher Electric Multipoles

Abstract Employing the Craig-Power Hamiltonian, a general formula is obtained for the retarded dispersion energy shift among three arbitrarily electrically polarisable atoms or molecules. Explicit expressions are then extracted for the first few leading order corrections to the triple dipole potential. These include energy shifts in which two of the particles are electric dipole polarisable and the third is either electric quadrupole or octupole polarisable, as well as the interaction between an electric dipole polarisable atom and two electric quadrupole polarisable ones. Results are obtained for arbitrary three-body configurations, and for equilateral triangle and collinear arrangements. In each case, near- and far-zone asymptotes are computed.

Keywords Generalised energy shift · Higher electric multipoles · Dipole-dipole-quadrupole · Dipole-quadrupole-quadrupole · Dipole-dipole-octupole

6.1 Introduction

One way in which the quality and accuracy of global potential energy surfaces may be improved upon, is by systematically including contributions from higher multipole moment correction terms. This is the case for properties that are particularly sensitive to forces at long-range arising from the electrostatic, induction and dispersion energy contributions. As shown in Chap. 4, interesting effects arise on accounting for magnetic dipole and electric octupole couplings. With collisional processes involving three particles, and interactions among many neutral non-polar species being dependent in a significant way on non-pairwise additive contributions to the van der Waals dispersion potential, it is important to go beyond the triple dipole energy shift and consider higher multipole terms. Here we restrict treatment to electric quadrupole and electric octupole interaction contributions, as these terms are expected to be dominant and potentially more applicable to molecular systems. Extension to include magnetic effects is not expected to pose any conceptual or computational difficulty.

In Sect. 5.3 it was seen how adoption of the Craig-Power Hamiltonian greatly simplified the calculation of the retarded analogue of the Axilrod-Teller-Muto

© The Author(s) 2016

A. Salam, *Non-Relativistic QED Theory of the van der Waals Dispersion Interaction*, SpringerBriefs in Electrical and Magnetic Properties of Atoms, Molecules, and Clusters, DOI 10.1007/978-3-319-45606-5_6

89

three-body dispersion potential. Only six time-ordered diagrams were required to be summed over at third-order of perturbation theory. Meanwhile in Sect. 4.2 it was shown how a generalised formula for the dispersion interaction between two molecules with arbitrary electric multipole polarisability could be easily obtained from knowledge of the functional form of the Casimir-Polder potential and be used to readily extract explicit higher electric contributions associated with quadrupole and octupole couplings. Here we combine these two approaches, and the inherent advantages offered by them, to derive a generalised expression for the pure electric multipole moment contribution of arbitrary order to the three-body retarded dispersion energy shift. The resulting formula is then used in subsequent sections to extract specific higher electric multipole contributions such as the electric dipole-dipole-quadrupole, dipole-dipole-octupole, etc. interaction energies, both for arbitrary arrangement of the three objects as well as for particular geometries. In the past these terms have been evaluated using semi-classical theory and have therefore only been applicable in the near-zone [1].

6.2 Generalised Three-Body Dispersion Potential

In this Section we derive a generalised expression for the retarded non-additive three-body dispersion energy shift between species possessing electric multipole polarisability characteristics of arbitrary order. Since symmetry prevents two different electric multipole moment operators from coupling the same virtual electronic state to the ground state in atoms, no contributions to the mixed electric multipole polarisability tensor are considered. An l-th order generalised polarisability at the real frequency ω is defined similarly to Eq. (4.8) as

$$\alpha^{(l)}_{i_1...i_l j_1...j_l}(\xi; \omega) = \sum_t \left\{ \frac{[E^{(l)}_{i_1...i_l}(\xi)]^{0r}[E^{(l)}_{j_1...j_l}(\xi)]^{r0}}{E^\xi_{r0} - \hbar\omega} + \frac{[E^{(l)}_{j_1...j_l}(\xi)]^{0r}[E^{(l)}_{i_1...i_l}(\xi)]^{r0}}{E^\xi_{r0} + \hbar\omega} \right\}, \qquad (6.1)$$

where the l-th order electric multipole moment operator $E^{(l)}_{i_1...i_l}(\xi)$ was defined by Eq. (4.1). The generalised form of the Craig-Power Hamiltonian, which in the electric dipole approximation was given by Eq. (5.8), is then straightforwardly written as [2]

$$H^{(l)}_{int} = -\frac{1}{2\varepsilon_0^2} \sum_{modes} \alpha^{(l)}_{i_1...i_l j_1...j_l}(\xi; k) \nabla_{i_2}...\nabla_{i_l} d^\perp_{i_1}(\vec{R}_\xi) \nabla_{j_2}...\nabla_{j_l} d^\perp_{j_1}(\vec{R}_\xi). \qquad (6.2)$$

Identical conditions on the Cartesian tensor components that led to the vanishing of the generalised interaction Hamiltonian linear in the electric displacement field Eq. (4.2) also apply to coupling Hamiltonian Eq. (6.2), namely that $H^{(l)}_{int} = 0$ when indices i_1 and j_1 associated with $\vec{d}^\perp(\vec{r})$ is equal to a subscript associated with any of the gradient operators appearing before the field.

Employing the generic form of Eq. (6.2) to three species A, B and C, characterised by an electrical polarisability of arbitrary order l, m and n, respectively, the computation of the dispersion potential follows identically to that presented in Sect. 5.3 for the triple dipole case. Again all three particles are in the electronic ground state, with excitations due to fluctuations of the vacuum electromagnetic field allowing virtual levels r, s and t in the three molecules, respectively, to be accessed. Six time-ordered graphs, one of which was illustrated in Fig. 5.1, depicting exchange of two virtual photons between any coupled material pair, contributes to the energy shift, with interaction vertices given by suitably index and multipole order modified forms of Eq. (6.2). Summing over the diagrams, third-order perturbation theory formula Eq. (5.9) gives for the generalised interaction energy

$$
\begin{aligned}
\Delta E^{(l,m,n)} = -\frac{2\hbar c}{(8\pi^2\varepsilon_0)^3} & (-\nabla^2\delta_{k_1 m_1} + \nabla_{k_1}\nabla_{m_1})^a (\nabla_{k_2}\ldots\nabla_{k_m}\nabla_{m_2}\ldots\nabla_{m_n})^a \\
& \times (-\nabla^2\delta_{i_1 n_1} + \nabla_{i_1}\nabla_{n_1})^b (\nabla_{i_2}\ldots\nabla_{i_l}\nabla_{n_2}\ldots\nabla_{n_n})^b \\
& \times (-\nabla^2\delta_{j_1 l_1} + \nabla_{j_1}\nabla_{l_1})^c (\nabla_{j_2}\ldots\nabla_{j_l}\nabla_{l_2}\ldots\nabla_{l_m})^c \frac{1}{abc} \\
& \times \int_0^\infty \int_0^\infty \int_0^\infty dk_1 dk_2 dk_3 \left[\frac{1}{(k_1+k_2)(k_2+k_3)} + \frac{1}{(k_1+k_3)(k_2+k_3)} + \frac{1}{(k_1+k_2)(k_1+k_3)} \right] \\
& \times [\alpha^{(l)}_{i_1\ldots i_l j_1\ldots j_l}(A;k_2) + \alpha^{(l)}_{i_1\ldots i_l j_1\ldots j_l}(A;k_3)][\alpha^{(m)}_{k_1\ldots k_m l_1\ldots l_m}(B;k_1) + \alpha^{(m)}_{k_1\ldots k_m l_1\ldots l_m}(B;k_3)] \\
& \times [\alpha^{(n)}_{m_1\ldots m_n n_1\ldots n_n}(C;k_1) + \alpha^{(n)}_{m_1\ldots m_n n_1\ldots n_n}(C;k_2)] \sin k_1 a \sin k_2 b \sin k_3 c,
\end{aligned}
$$

$$(6.3)$$

where the distances a, b and c were defined in Sect. 5.3 appropriate to a triangular configuration of the three particles. In arriving at Eq. (6.3), mode polarisation and wave vector sums have been carried out according to identities (3.18) and (3.19), respectively, while relation (5.11) was used to perform the angular average. Rewriting the sum of energy denominators in terms of the parameter u via the integral relation Eq. (5.12), and carrying out the ensuing complex integrals using the result Eq. (5.13) [3], yields the general formula [2]

$$
\begin{aligned}
\Delta E^{(l,m,n)} = -\frac{\hbar c}{64\pi^4\varepsilon_0^3} & (-\nabla^2\delta_{k_1 m_1} + \nabla_{k_1}\nabla_{m_1})^a (\nabla_{k_2}\ldots\nabla_{k_m}\nabla_{m_2}\ldots\nabla_{m_n})^a \\
& \times (-\nabla^2\delta_{i_1 n_1} + \nabla_{i_1}\nabla_{n_1})^b (\nabla_{i_2}\ldots\nabla_{i_l}\nabla_{n_2}\ldots\nabla_{n_n})^b \\
& \times (-\nabla^2\delta_{j_1 l_1} + \nabla_{j_1}\nabla_{l_1})^c (\nabla_{j_2}\ldots\nabla_{j_l}\nabla_{l_2}\ldots\nabla_{l_m})^c \frac{1}{abc} \\
& \times \int_0^\infty du\, \alpha^{(l)}_{i_1\ldots i_l j_1\ldots j_l}(A;iu)\alpha^{(m)}_{k_1\ldots k_m l_1\ldots l_m}(B;iu)\alpha^{(n)}_{m_1\ldots m_n n_1\ldots n_n}(C;iu)e^{-u(a+b+c)},
\end{aligned}
$$

$$(6.4)$$

for species A which is pure l-th order electric polarisable, B which is m-th order electric polarisable, and n-th order C, with the generalised polarisability tensors evaluated at imaginary frequency $\omega = icu$ as before. Expression (6.4) holds for all separation distances a, b and c beyond orbital overlap and extending out to infinity, for oriented A, B and C. It correctly includes the effects of retardation. On inserting $l = m = n = 1$ into Eq. (6.4), the retarded triple dipole dispersion potential Eq. (5.14) is found as expected, with $\alpha_{ij}^{(1)}(\xi; iu)$ easily obtainable from definition (6.1) or from (4.8).

6.3 Dipole-Dipole-Quadrupole Potential

The leading correction involving pure electric multipole moment polarisabilities to the Aub-Zienau dispersion energy arising between three electric dipole polarisable bodies is obtained by characterising one of the species to be electric quadrupole polarisable. Let this be particle C. For oriented A, B and C, the dispersion interaction energy follows straightforwardly from the generalised formula (6.4) on substituting $l = m = 1$ and $n = 2$. Whence [2],

$$\Delta E^{(1,1,2)} = -\frac{\hbar c}{64\pi^4 \varepsilon_0^3}(-\nabla^2 \delta_{km_1} + \nabla_k \nabla_{m_1})^a \nabla_{m_2}^a(-\nabla^2 \delta_{in_1} + \nabla_i \nabla_{n_1})^b \nabla_{n_2}^b(-\nabla^2 \delta_{jl} + \nabla_j \nabla_l)^c \frac{1}{abc}$$

$$\times \int_0^\infty du \alpha_{ij}^{(1)}(A; iu)\alpha_{kl}^{(1)}(B; iu)\alpha_{m_1 m_2 n_1 n_2}^{(2)}(C; iu)e^{-u(a+b+c)},$$

$$(6.5)$$

where the pure electric quadrupole polarisability at imaginary wave vector of object C, $\alpha_{m_1 m_2 n_1 n_2}^{(2)}(C; iu)$, is obtainable from definition (6.1) or as given in Eq. (4.12). The transition electric quadrupole moment $[E_{m_1 m_2}^{(2)}(C)]^{0t}$ featuring in this last mentioned quantity is again traceless, symmetric in Cartesian tensor components, and of irreducible weight-2.

To obtain the potential for isotropic bodies, an orientational average is carried out. As previously, this may be done as separate averages over each particle using earlier results: relation (3.27) for electric dipole dependent terms, and Eq. (4.14) for electric quadrupole polarisability. On making use of the geometric tensors $F_{ij}(uR)$, Eq. (5.15), and $H_{ijk}(uR)$, Eq. (4.11), the energy shift for randomly averaged particles may be written compactly as

$$\Delta E^{(1,1,2)} = -\frac{\hbar c}{64\pi^4 \varepsilon_0^3} \int_0^\infty du \alpha^{(1)}(A; iu)\alpha^{(1)}(B; iu)\alpha_{\lambda\mu\lambda\mu}^{(2)}(C; iu)H_{jkl}(ua)$$

$$\times [H_{ikl}(ub) + H_{ilk}(ub)]F_{ij}(uc). \qquad (6.6)$$

An expression for $\Delta E^{(1,1,2)}$ explicitly in terms of direction cosines involving unit vectors $\hat{a}, \hat{b}$ and $\hat{c}$ is obtained after multiplying the H_{ijk} and F_{ij} factors, and which is applicable for an arbitrary triangular configuration of the three particles, is given in Ref. [2]. Note that a factor of 1/3 has been absorbed into each of the dipole polarisabilities, and a factor of 1/10 into the isotropic quadrupole polarisability.

The near-zone limiting form, independent of u, is short enough to be given below:

$$\Delta E_{NZ}^{(1,1,2)} = \frac{9\hbar cX}{32\pi^4 \varepsilon_0^3 a^4 b^4 c^3} \frac{1}{} \int_0^\infty du \alpha^{(1)}(A; iu)\alpha^{(1)}(B; iu)\alpha_{\lambda\mu\lambda\mu}^{(2)}(C; iu)$$

$$\times [-21(\hat{a} \cdot \hat{b}) + 3(\hat{b} \cdot \hat{c})(\hat{c} \cdot \hat{a}) + 25(\hat{a} \cdot \hat{b})^3 + 30(\hat{a} \cdot \hat{b})(\hat{c} \cdot \hat{a})^2$$
$$+ 30(\hat{a} \cdot \hat{b})(\hat{b} \cdot \hat{c})^2 - 75(\hat{a} \cdot \hat{b})^2(\hat{b} \cdot \hat{c})(\hat{c} \cdot \hat{a})]. \quad (6.7)$$

Result (6.7) is identical to that first obtained by Bell [1], who used static dipolar and quadrupolar coupling potentials. It exhibits inverse quartic dependence on separations a and b, and inverse cubic power law with respect to distance variable c.

Adopting an equilateral triangle arrangement, for which $a = b = c = R$ as before, and setting internal angles to be $60°$, the retarded dispersion potential (6.6) becomes [4]

$$\Delta E_{Eq}^{(1,1,2)} = -\frac{\hbar c}{512\pi^4 \varepsilon_0^3 R^{11}} \int_0^\infty du \alpha^{(1)}(A; iu)\alpha^{(1)}(B; iu)\alpha_{\lambda\mu\lambda\mu}^{(2)}(C; iu)e^{-3uR}$$

$$\times [5u^8 R^8 + 23u^7 R^7 + 83u^6 R^6 + 270u^5 R^5 + 717u^4 R^4$$
$$+ 1473u^3 R^3 + 2106u^2 R^2 + 1755uR + 585]. \quad (6.8)$$

With $uR \gg 1$ in the far-zone, approximating the polarisabilities by their static forms and evaluating the u-integrals yields an R^{-12} asymptotic limit

$$\Delta E_{FZ}^{(1,1,2)} = -\frac{22567 \times 5\hbar c}{2^5 \times 3^7 \pi^4 \varepsilon_0^3 R^{12}}\alpha^{(1)}(A; 0)\alpha^{(1)}(B; 0)\alpha_{\lambda\mu\lambda\mu}^{(2)}(C; 0). \quad (6.9)$$

In contrast, the near-zone energy shift displays an R^{-11} dependence. It may be obtained directly from Eq. (6.7) on substituting parameters applicable to an equilateral triangle, or from (6.8) on letting $uR \ll 1$, and keeping the factor of 585 from the terms within square brackets. Thus

$$\Delta E_{NZ}^{(1,1,2)} = -\frac{585\hbar c}{512\pi^4 \varepsilon_0^3 R^{11}} \int_0^\infty du \alpha^{(1)}(A; iu)\alpha^{(1)}(B; iu)\alpha_{\lambda\mu\lambda\mu}^{(2)}(C; iu). \quad (6.10)$$

Explicitly in terms of electric multipole moments, (6.10) may be written alternatively as

$$\Delta E_{NZ}^{(1,1,2)} = -\frac{13}{256\pi^3\varepsilon_0^3 R^{11}} \sum_{r,s,t} \left|[\vec{E}^{(1)}(A)]^{0r}\right|^2 \left|[\vec{E}^{(1)}(B)]^{0s}\right|^2 [E_{\lambda\mu}^{(2)}(C)]^{0t}[E_{\lambda\mu}^{(2)}(C)]^{t0}$$

$$\times \frac{(E_{r0}+E_{s0}+E_{t0})}{(E_{r0}+E_{s0})(E_{s0}+E_{t0})(E_{t0}+E_{r0})},$$

(6.11)

on employing integral relation (5.21). For this geometry, the interaction energy and its asymptotic limits are all attractive.

On inserting the geometrical parameters appropriate to all three particles lying on one line, with C placed in the middle, Eq. (6.6) yields the dispersion potential [4]

$$\Delta E_{Coll}^{(1,1,2)} = \frac{\hbar c}{8\pi^4\varepsilon_0^3 R^{11}} \int_0^\infty du\, \alpha^{(1)}(A;iu)\alpha^{(1)}(B;iu)\alpha_{\lambda\mu\lambda\mu}^{(2)}(C;iu)e^{-2uR}$$

$$\times [u^8 R^8 + 13u^7 R^7 + 97u^6 R^6 + 432u^5 R^5 + 996u^4 R^4$$

$$+ 384u^3 R^3 - 4032u^2 R^2 - 9216uR - 4608].$$

(6.12)

It is worth remarking that for a collinear arrangement of the three bodies, in which B is electric quadrupole polarisable, with C still in between A and B, the energy shift is identical to result (6.12). Hence $\Delta E_{Coll}^{(1,2,1)} = \Delta E_{Coll}^{(1,1,2)}$.

Retaining the u-independent term within square brackets of Eq. (6.12), and letting $e^{-2uR} \to 1$ yields the short-range asymptote

$$\Delta E_{NZ}^{(1,1,2)} = -\frac{576\hbar c}{\pi^4\varepsilon_0^3 R^{11}} \int_0^\infty du\, \alpha^{(1)}(A;iu)\alpha^{(1)}(B;iu)\alpha_{\lambda\mu\lambda\mu}^{(2)}(C;iu)$$

$$= -\frac{128}{5\pi^3\varepsilon_0^3 R^{11}} \sum_{r,s,t} \left|[\vec{E}^{(1)}(A)]^{0r}\right|^2 \left|[\vec{E}^{(1)}(B)]^{0s}\right|^2 [E_{\lambda\mu}^{(2)}(C)]^{0t}[E_{\lambda\mu}^{(2)}(C)]^{t0}$$

$$\times \frac{(E_{r0}+E_{s0}+E_{t0})}{(E_{r0}+E_{s0})(E_{s0}+E_{t0})(E_{t0}+E_{r0})},$$

(6.13)

while at very long-range, Eq. (6.12) reduces to

$$\Delta E_{FZ}^{(1,1,2)} = -\frac{48555\hbar c}{128\pi^4\varepsilon_0^3 R^{12}}\alpha^{(1)}(A;0)\alpha^{(1)}(B;0)\alpha_{\lambda\mu\lambda\mu}^{(2)}(C;0),$$

(6.14)

exhibiting identical respective power law dependences to their equilateral triangle counterparts. Both asymptotes (6.13) and (6.14) are also attractive.

6.4 Dipole-Quadrupole-Quadrupole Dispersion Energy Shift

The next higher-order contribution to the three-body dispersion potential involving species with pure electric multipole polarisability characteristics is between an electric dipole polarisable molecule, A and two objects with electric quadrupole polarisability, B and C. Inserting $l = 1$ and $m = n = 2$ into the general formula Eq. (6.4) allows the interaction energy for oriented particles to be written down directly as

$$
\Delta E^{(1,2,2)} = -\frac{\hbar c}{64\pi^4 \varepsilon_0^3}(-\nabla^2 \delta_{k_1 m_1} + \nabla_{k_1}\nabla_{m_1})^a \nabla_{k_2}^a \nabla_{m_2}^a (-\nabla^2 \delta_{in_1} + \nabla_i \nabla_{n_1})^b \nabla_{n_2}^b
$$

$$
\times (-\nabla^2 \delta_{jl_1} + \nabla_j \nabla_{l_1})^c \nabla_{l_2}^c \frac{1}{abc} \int_0^{\infty} due^{-u(a+b+c)} \alpha_{ij}^{(1)}(A; iu)\alpha_{k_1 k_2 l_1 l_2}^{(2)}(B; iu)\alpha_{m_1 m_2 n_1 n_2}^{(2)}(C; iu).
$$

$$(6.15)$$

In terms of the tensors $H_{ijk}(uR)$, Eq. (4.11), and $L_{ijkl}(uR)$, Eq. (4.22), Eq. (6.15) becomes

$$
\Delta E^{(1,2,2)} = -\frac{\hbar c}{64\pi^4 \varepsilon_0^3}\int_0^{\infty} du \alpha_{ij}^{(1)}(A; iu)\alpha_{k_1 k_2 l_1 l_2}^{(2)}(B; iu)\alpha_{m_1 m_2 n_1 n_2}^{(2)}(C; iu)
$$

$$
\times L_{k_1 m_1 k_2 m_2}(ua) H_{in_1 n_2}(ub) H_{jl_1 l_2}(uc).
$$

$$(6.16)$$

Orientationally averaging $\alpha_{ij}^{(1)}(A; iu)$ using Eq. (3.27), and the quadrupole polarisabilities via Eq. (4.14), produces for isotropic bodies, the energy shift

$$
\Delta E^{(1,2,2)} = -\frac{\hbar c}{64\pi^4 \varepsilon_0^3}\int_0^{\infty} du \alpha^{(1)}(A; iu)\alpha_{\lambda\mu\lambda\mu}^{(2)}(B; iu)\alpha_{\nu\pi\nu\pi}^{(2)}(C; iu)
$$

$$
\times L_{jlkm}(ua)[H_{ilm}(ub) + H_{iml}(ub)][H_{ijk}(uc) + H_{ikj}(uc)].
$$

$$(6.17)$$

Multiplying the L_{ijkl} and H_{ijk} factors yields a lengthy expression explicitly in terms of the direction cosines $\hat{a} \cdot \hat{b}$, $\hat{b} \cdot \hat{c}$ and $\hat{c} \cdot \hat{a}$, and which may be found in Ref. [2]. The near-zone asymptote for a triangular arrangement with arbitrary side lengths is brief enough to be given here. It has the form

$$\Delta E_{NZ}^{(1,2,2)} = -\frac{27\hbar c}{64\pi^4\varepsilon_0^3 a^5 b^4 c^4}\int\limits_0^\infty du\,\alpha^{(1)}(A;iu)\alpha_{\lambda\mu\lambda\mu}^{(2)}(B;iu)\alpha_{\nu\pi\nu\pi}^{(2)}(C;iu)$$

$$\times\,[-140(\hat{b}\cdot\hat{c}) + 1200(\hat{a}\cdot\hat{b})(\hat{c}\cdot\hat{a}) + 100(\hat{c}\cdot\hat{a})^2(\hat{b}\cdot\hat{c})$$
$$+\,100(\hat{a}\cdot\hat{b})^2(\hat{b}\cdot\hat{c}) + 200(\hat{b}\cdot\hat{c})^3$$
$$-\,1400(\hat{c}\cdot\hat{a})^3(\hat{a}\cdot\hat{b}) - 1400(\hat{a}\cdot\hat{b})^3(\hat{c}\cdot\hat{a}) - 2000(\hat{b}\cdot\hat{c})^2(\hat{a}\cdot\hat{b})(\hat{c}\cdot\hat{a})$$
$$+\,3500(\hat{a}\cdot\hat{b})^2(\hat{c}\cdot\hat{a})^2(\hat{b}\cdot\hat{c})],$$

(6.18)

and which is equivalent to the result obtained using semi-classical theory [1]. The potential has inverse fifth power dependence on a, and inverse fourth power behaviour on each of b and c.

Specialising to an equilateral triangle configuration, with identical side length R, and equal internal angles, the energy shift is [4]

$$\Delta E_{Eq}^{(1,2,2)} = \frac{\hbar c}{512\pi^4\varepsilon_0^3 R^{13}}\int\limits_0^\infty du\,e^{-3uR}\alpha^{(1)}(A;iu)\alpha_{\lambda\mu\lambda\mu}^{(2)}(B;iu)\alpha_{\nu\pi\nu\pi}^{(2)}(C;iu)$$

$$\times\,[u^{10}R^{10} + 26u^9R^9 + 189u^8R^8 + 993u^7R^7$$
$$+\,4284u^6R^6 + 14562u^5R^5 + 36918u^4R^4$$
$$+\,66501u^3R^3 + 80082u^2R^2 + 57915uR + 19305].$$

(6.19)

Limiting forms result on making the familiar approximations. In the far-zone (6.19) tends to the asymptote

$$\Delta E_{FZ}^{(1,2,2)} = \frac{110991\hbar c}{3^7\times 8\pi^4\varepsilon_0^3 R^{14}}\alpha^{(1)}(A;0)\alpha_{\lambda\mu\lambda\mu}^{(2)}(B;0)\alpha_{\nu\pi\nu\pi}^{(2)}(C;0),$$

(6.20)

exhibiting R^{-14} behaviour, and is proportional to the static polarisabilities. The near-zone potential, which may be obtained from Eq. (6.19) on keeping the u-independent term and setting $e^{-3uR} = 1$, or directly from Eq. (6.18) on substituting parameters for a, b and c, and direction cosines applicable for an equilateral triangle, is found to be

$$\Delta E_{NZ}^{(1,2,2)} = \frac{19305\hbar c}{512\pi^4\varepsilon_0^3 R^{13}} \int_0^\infty du\, \alpha^{(1)}(A;iu)\alpha_{\lambda\mu\lambda\mu}^{(2)}(B;iu)\alpha_{\nu\pi\nu\pi}^{(2)}(C;iu)$$

$$= \frac{1287}{2560\pi^3\varepsilon_0^3 R^{13}} \sum_{r,s,t} \left|[\vec{E}^{(1)}(A)]^{0r}\right|^2 [E_{\lambda\mu}^{(2)}(B)]^{0s}[E_{\lambda\mu}^{(2)}(B)]^{s0}[E_{\nu\pi}^{(2)}(C)]^{0t}[E_{\nu\pi}^{(2)}(C)]^{t0}$$

$$\times \frac{(E_{r0}+E_{s0}+E_{t0})}{(E_{r0}+E_{s0})(E_{s0}+E_{t0})(E_{t0}+E_{r0})},$$

$$(6.21)$$

displaying R^{-13} dependence on separation distance. Note the potential has positive sign.

As for other three-body dispersion potentials considered thus far, results are given for a collinear geometry A-C-B. From Eq. (6.17), on choosing $a = b = c/2 = R/2$, and $\theta_A = \theta_B = 0°$, and $\theta_C = 180°$, the energy shift valid for all R outside wave function overlap is found to be [4]

$$\Delta E_{Coll}^{(1,2,2)} = -\frac{\hbar c}{8\pi^4\varepsilon_0^3 R^{13}} \int_0^\infty du\, e^{-2uR}\alpha^{(1)}(A;iu)\alpha_{\lambda\mu\lambda\mu}^{(2)}(B;iu)\alpha_{\nu\pi\nu\pi}^{(2)}(C;iu)$$

$$\times [u^{10}R^{10} + 19u^9 R^9 + 222u^8 R^8 + 1638u^7 R^7$$

$$+ 7464u^6 R^6 + 18000u^5 R^5 - 432u^4 R^4$$

$$- 147456u^3 R^3 - 442368u^2 R^2 - 552960uR - 276480]. \qquad (6.22)$$

Expression (6.22) also applies to the situation in which the electric dipole polarisable molecule lies mid-way between the two electric quadrupole polarisable species. At very long-range, result (6.22) tends to the limit

$$\Delta E_{FZ}^{(1,2,2)} = \frac{2469771\hbar c}{64\pi^4\varepsilon_0^3 R^{14}}\alpha^{(1)}(A;0)\alpha_{\lambda\mu\lambda\mu}^{(2)}(B;0)\alpha_{\nu\pi\nu\pi}^{(2)}(C;0), \qquad (6.23)$$

which has repulsive R^{-14} behaviour. In the near-zone the interaction energy is

$$\Delta E_{NZ}^{(1,2,2)} = \frac{34560\hbar c}{\pi^4\varepsilon_0^3 R^{13}} \int_0^\infty du\, \alpha^{(1)}(A;iu)\alpha_{\lambda\mu\lambda\mu}^{(2)}(B;iu)\alpha_{\nu\pi\nu\pi}^{(2)}(C;iu)$$

$$= \frac{2304}{5\pi^3\varepsilon_0^3 R^{13}} \sum_{r,s,t} \left|[\vec{E}^{(1)}(A)]^{0r}\right|^2 [E_{\lambda\mu}^{(2)}(B)]^{0s}[E_{\lambda\mu}^{(2)}(B)]^{s0}[E_{\nu\pi}^{(2)}(C)]^{0t}[E_{\nu\pi}^{(2)}(C)]^{t0}$$

$$\times \frac{(E_{r0}+E_{s0}+E_{t0})}{(E_{r0}+E_{s0})(E_{s0}+E_{t0})(E_{t0}+E_{r0})},$$

$$(6.24)$$

which is stronger by a factor of R relative to the far-zone asymptote (6.23).

6.5   Dipole-Dipole-Octupole Dispersion Potential

The last higher-order electric multipole contribution to the three-body dispersion energy shift to be considered is that between two species A and B which are electric dipole polarisable, and a third molecule, C, which is electric octupole polarisable. From the general formula (6.4), the interaction energy between oriented particles is easily obtained on letting $l = m = 1$ and $n = 3$. Thus

$$\Delta E^{(1,1,3)} = -\frac{\hbar c}{64\pi^4 \varepsilon_0^3}(-\nabla^2 \delta_{km_1} + \nabla_k \nabla_{m_1})^a (\nabla_{m_2} \nabla_{m_3})^a (-\nabla^2 \delta_{in_1} + \nabla_i \nabla_{n_1})^b (\nabla_{n_2} \nabla_{n_3})^b$$

$$\times (-\nabla^2 \delta_{jl} + \nabla_j \nabla_l)^c \frac{1}{abc} \int_0^\infty du e^{-u(a+b+c)} \alpha_{ij}^{(1)}(A; iu)\alpha_{kl}^{(1)}(B; iu)\alpha_{m_1 m_2 m_3 n_1 n_2 n_3}^{(3)}(C; iu),$$

$$(6.25)$$

where the pure electric octupole polarisability tensor of species C, $\alpha_{m_1 m_2 m_3 n_1 n_2 n_3}^{(3)}(C; iu)$ at imaginary wave vector $k = iu$, was defined in terms of reducible components of the electric octupole moment, $E_{m_1 m_2 m_3}^{(3)}(C)$, Eq. (4.28), by relation (4.27). Utilising the $F_{ij}(uR)$ and $L_{ijkl}(uR)$ tensors, the potential may be written as [2, 4]

$$\Delta E^{(1,1,3)} = -\frac{\hbar c}{64\pi^4 \varepsilon_0^3} \int_0^\infty du \alpha_{ij}^{(1)}(A; iu)\alpha_{kl}^{(1)}(B; iu)\alpha_{m_1 m_2 m_3 n_1 n_2 n_3}^{(3)}(C; iu)$$

$$\times L_{km_1 m_2 m_3}(ua)L_{in_1 n_2 n_3}(ub)F_{jl}(uc). \qquad (6.26)$$

Independent averages of second and sixth rank Cartesian tensors for dipole and octupole polarisabilities, respectively, may be performed to yield the energy shift for freely tumbling entities. Retaining the coefficient 1/210 explicitly as a pre-factor in the isotropic octupole polarisability, the orientationally averaged potential energy is [2]

$$\Delta E^{(1,1,3)} = -\frac{\hbar c}{15 \times 64\pi^4 \varepsilon_0^3} \int_0^\infty du \alpha^{(1)}(A; iu)\alpha^{(1)}(B; iu)$$

$$\times \begin{pmatrix} \alpha_{\lambda\mu\mu\lambda\nu\nu}^{(3)}(C; iu)L_{jkll}(ua)L_{ikmm}(ub)F_{ij}(uc) \\ + 2\alpha_{\lambda\mu\nu\lambda\mu\nu}^{(3)}(C; iu)\{L_{jklm}(ua)[2L_{iklm}(ub) + L_{ilkm}(ub) \\ + L_{ilmk}(ub) + L_{imkl}(ub) + L_{imlk}(ub)]F_{ij}(uc)\} \end{pmatrix}, \qquad (6.27)$$

and which may be decomposed into contributions dependent upon octupole weight-1 and weight-3 terms on using relation (4.29) as done in Sect. 4.5 for the two-body electric dipole-octupole dispersion energy. Multiplying the geometric tensors in Eq. (6.27) produces a formula that depends on direction cosines formed from unit displacement vectors $\hat{a}, \hat{b}$, and $\hat{c}$, and distances a, b and c [2].

When A, B and C adopt an equilateral triangle geometry, the retarded electric dipole-dipole-octupole dispersion energy is

$$
\begin{aligned}
\Delta E_{Eq}^{(1,1,3)} = {} & \frac{\hbar c}{64 \times 120\pi^4 \varepsilon_0^3 R^{13}} \int_0^\infty du\, e^{-3uR} \alpha^{(1)}(A; iu)\alpha^{(1)}(B; iu)\alpha_{\lambda\mu\mu\lambda\nu\nu}^{(3)}(C; iu) \\
& \times [33u^4 R^4 + 99u^5 R^5 + 120u^6 R^6 + 75u^7 R^7 + 24u^8 R^8 + 3u^9 R^9 + 7u^{10} R^{10}] \\
& + \frac{\hbar c}{64 \times 120\pi^4 \varepsilon_0^3 R^{13}} \int_0^\infty du\, e^{-3uR} \alpha^{(1)}(A; iu)\alpha^{(1)}(B; iu)\alpha_{\lambda\mu\nu\lambda\mu\nu}^{(3)}(C; iu) \\
& \times [13u^{10} R^{10} + 119u^9 R^9 + 785u^8 R^8 + 2784u^7 R^7 + 5307u^6 R^6 + 3789u^5 R^5 \\
& - 1446u^4 R^4 + 9441u^3 R^3 + 46332u^2 R^2 + 58725uR + 19575],
\end{aligned}
$$

$$(6.28)$$

when partitioned into a sum of two terms, one proportional to $E_{\lambda\mu\mu}^{(3)}(C)$ and the other dependent upon $E_{\lambda\mu\nu}^{(3)}(C)$ contributions.

Comparing the first term of (6.28) with the triple electric dipole potential for the same geometry, Eq. (5.26), it is seen that the coefficients preceding each polynomial term in uR is identical in both cases, noting the additional $(uR)^4$ factor in the octupole case since coupling of this multipole moment to the electric displacement field involves the action of two gradient operators. As in the analogous two-body example, the first term of (6.28), involving weight-1 octupole polarisability is a higher-order correction to the retarded triple dipole dispersion potential of Aub and Zienau [5]. This feature holds true for arbitrary triangular geometry, and will be shown explicitly below for the collinear configuration of three objects.

Since there is no term independent of u in the first integral of (6.28), the near-zone potential is obtained by keeping the term with coefficient 19,575 within square brackets of the second integral, while simultaneously letting $e^{-3uR} \approx 1$, to give at short-range, but still outside orbital overlap, the energy shift formula

$$
\Delta E_{NZ}^{(1,1,3)} = \frac{1305\hbar c}{512\pi^4 \varepsilon_0^3 R^{13}} \int_0^\infty du\, \alpha^{(1)}(A; iu)\alpha^{(1)}(B; iu)\alpha_{\lambda\mu\nu\lambda\mu\nu}^{(3)}(C; iu), \tag{6.29}
$$

which like the near-zone dipole-quadrupole-quadrupole potential, has R^{-13} power law dependence. Both terms of (6.28) contribute in the far-zone. Evaluating the u-integrals yields

$$\Delta E_{FZ}^{(1,1,3)} = \frac{\hbar c}{15 \times 2^9 \times 3^7 \pi^4 \varepsilon_0^3 R^{14}} \alpha^{(1)}(A;0)\alpha^{(1)}(B;0)$$
$$\times \left[716608\alpha_{\lambda\mu\mu\lambda\nu\nu}^{(3)}(C;0) + 52824000\alpha_{\lambda\mu\nu\lambda\mu\nu}^{(3)}(C;0)\right], \qquad (6.30)$$

exhibiting R^{-14} dependence.

When the three objects lie along a straight line, with A and B either side of C, the interaction energy is [4]

$$\Delta E_{Coll}^{(1,1,3)} = \frac{\hbar c}{120\pi^4 \varepsilon_0^3 R^{13}} \int_0^\infty du\, e^{-2uR} \alpha^{(1)}(A;iu)\alpha^{(1)}(B;iu)\alpha_{\lambda\mu\mu\lambda\nu\nu}^{(3)}(C;iu)$$

$$\times \left[-48u^4 R^4 - 96u^5 R^5 - 36u^6 R^6 + 16u^7 R^7 + 17u^8 R^8 + 5u^9 R^9 + u^{10} R^{10}\right]$$

$$+ \frac{\hbar c}{30\pi^4 \varepsilon_0^3 R^{13}} \int_0^\infty du\, e^{-2uR} \alpha^{(1)}(A;iu)\alpha^{(1)}(B;iu)\alpha_{\lambda\mu\nu\lambda\mu\nu}^{(3)}(C;iu)[u^{10} R^{10} + 21u^9 R^9$$

$$+ 273u^8 R^8 + 2498u^7 R^7 + 14790u^6 R^6 + 352880u^5 R^5$$

$$+ 127576u^4 R^4 + 50688u^3 R^3 - 57344u^2 R^2 - 1382400uR - 691200]$$
$$(6.31)$$

Inspection of the triple dipole dispersion potential for collinear geometry Eq. (5.30) reveals that the first term of (6.31) may be viewed as a higher-order correction to the former energy shift due to identical structural features of the term, and given that it is proportional to octupole weight-1, which has the characteristics of a vector. Because the first integral term of (6.31) only contributes to the retarded potential, the near-zone limit is obtained from the second integral term. It is found to be

$$\Delta E_{NZ}^{(1,1,3)} = -\frac{23040\hbar c}{\pi^4 \varepsilon_0^3 R^{13}} \int_0^\infty du\, \alpha^{(1)}(A;iu)\alpha^{(1)}(B;iu)\alpha_{\lambda\mu\nu\lambda\mu\nu}^{(3)}(C;iu)$$

$$= -\frac{10240}{\pi^3 \varepsilon_0^3 R^{13}} \sum_{r,s,t} |[\vec{E}^{(1)}(A)]^{0r}|^2 |[\vec{E}^{(1)}(B)]^{0s}|^2 [E_{\lambda\mu\nu}^{(3)}(C)]^{0t}[E_{\lambda\mu\nu}^{(3)}(C)]^{t0}$$

$$\times \frac{(E_{r0} + E_{s0} + E_{t0})}{(E_{r0} + E_{s0})(E_{s0} + E_{t0})(E_{t0} + E_{r0})}, \qquad (6.32)$$

with attractive R^{-13} dependence. In the far-zone, Eq. (6.31) tends to the limit

$$\Delta E_{FZ}^{(1,1,3)} = \frac{\hbar c}{\pi^4 \varepsilon_0^3 R^{14}} \alpha^{(1)}(A;0)\alpha^{(1)}(B;0)\left[\frac{1593}{40}\alpha_{\lambda\mu\mu\lambda\nu\nu}^{(3)}(C;0) + \frac{1871117}{240}\alpha_{\lambda\mu\nu\lambda\mu\nu}^{(3)}(C;0)\right],$$
$$(6.33)$$

the energy shift depending on both $E^{(3)}_{\lambda\mu\mu}(C)$ and $E^{(3)}_{\lambda\mu\nu}(C)$ contributions to the electric octupole moment, and displaying R^{-14} behaviour.

Finally, it should be pointed out that higher-order corrections to the triple dipole dispersion potential occur when one of the electric dipole moments in each species is successively replaced by an electric octupole moment, to yield DD-DD-DO, DD-DO-DO, and DO-DO-DO potentials, for example. In these cases the contribution from the octupole weight-3 dependent term vanishes on orientational averaging, leaving a fully retarded contribution that depends solely on octupole weight-1 moment. For the DD-DO-OO interaction energy, the pure electric octupole polarisability contains both octupole weight-1 and weight-3 dependent terms. In all of the cases involving electric multipole coupling, retardation effects weaken the potential.

References

1. Bell RJ (1970) Multipolar expansion for the non-additive third-order interaction energy of three atoms. J Phys B: At Mol Phys 3:751
2. Salam A (2014) Dispersion potential between three bodies with arbitrary electric multipole polarisabilities: molecular QED theory. J Chem Phys 140:044111
3. Passante R, Power EA, Thirunamachandran T (1998) Radiation-molecule coupling using dynamic polarisabilities: application to many-body forces. Phys Lett A 249:77
4. Salam A (2013) Higher-order electric multipole contributions to retarded non-additive three-body dispersion interaction energies between atoms: equilateral triangle and collinear configurations. J Chem Phys 139:244105
5. Aub MR, Zienau S (1960) Studies on the retarded interaction between neutral atoms. I. Three-body London-van der Waals interaction of neutral atoms. Proc Roy Soc London A257:464

Index

© The Author(s) 2016
A. Salam, *Non-Relativistic QED Theory of the van der Waals
Dispersion Interaction*, SpringerBriefs in Electrical and Magnetic Properties
of Atoms, Molecules, and Clusters, DOI 10.1007/978-3-319-45606-5

Printed in the United States
By Bookmasters

Printed in the United States
By Bookmasters